Theoretical Physics 5

T0171836

Wolfgang Nolting

Theoretical Physics 5

Thermodynamics

 Springer

Wolfgang Nolting
Inst. Physik
Humboldt-Universität zu Berlin
Berlin, Germany

ISBN 978-3-319-83855-7   ISBN 978-3-319-47910-1   (eBook)
DOI 10.1007/978-3-319-47910-1

© Springer International Publishing AG 2017
Softcover reprint of the hardcover 1st edition 2017
This work is subject to copyright. All rights are reserved by the Publisher, whether the whole or part of
the material is concerned, specifically the rights of translation, reprinting, reuse of illustrations, recitation,
broadcasting, reproduction on microfilms or in any other physical way, and transmission or information
storage and retrieval, electronic adaptation, computer software, or by similar or dissimilar methodology
now known or hereafter developed.
The use of general descriptive names, registered names, trademarks, service marks, etc. in this publication
does not imply, even in the absence of a specific statement, that such names are exempt from the relevant
protective laws and regulations and therefore free for general use.
The publisher, the authors and the editors are safe to assume that the advice and information in this book
are believed to be true and accurate at the date of publication. Neither the publisher nor the authors or
the editors give a warranty, express or implied, with respect to the material contained herein or for any
errors or omissions that may have been made.

Printed on acid-free paper

This Springer imprint is published by Springer Nature
The registered company is Springer International Publishing AG
The registered company address is: Gewerbestrasse 11, 6330 Cham, Switzerland

General Preface

The nine volumes of the series *Basic Course: Theoretical Physics* are thought to be text book material for the study of university level physics. They are aimed to impart, in a compact form, the most important skills of theoretical physics which can be used as basis for handling more sophisticated topics and problems in the advanced study of physics as well as in the subsequent physics research. The conceptual design of the presentation is organized in such a way that

Classical Mechanics (volume 1)
Analytical Mechanics (volume 2)
Electrodynamics (volume 3)
Special Theory of Relativity (volume 4)
Thermodynamics (volume 5)

are considered as the theory part of an *integrated course* of experimental and theoretical physics as is being offered at many universities starting from the first semester. Therefore, the presentation is consciously chosen to be very elaborate and self-contained, sometimes surely at the cost of certain elegance, so that the course is suitable even for self-study, at first without any need of secondary literature. At any stage, no material is used which has not been dealt with earlier in the text. This holds in particular for the mathematical tools, which have been comprehensively developed starting from the school level, of course more or less in the form of recipes, such that right from the beginning of the study, one can solve problems in theoretical physics. The mathematical insertions are always then plugged in when they become indispensable to proceed further in the program of theoretical physics. It goes without saying that in such a context, not all the mathematical statements can be proved and derived with absolute rigor. Instead, sometimes a reference must be made to an appropriate course in mathematics or to an advanced textbook in mathematics. Nevertheless, I have tried for a reasonably balanced representation so that the mathematical tools are not only applicable but also appear at least "plausible".

The mathematical interludes are of course necessary only in the first volumes of this series, which incorporate more or less the material of a bachelor program. In the second part of the series which comprises the modern aspects of theoretical physics,

Quantum Mechanics: Basics (volume 6)
Quantum Mechanics: Methods and Applications (volume 7)
Statistical Physics (volume 8)
Many-Body Theory (volume 9),

mathematical insertions are no longer necessary. This is partly because, by the time one comes to this stage, the obligatory mathematics courses one has to take in order to study physics would have provided the required tools. The fact that training in theory has already started in the first semester itself permits inclusion of parts of quantum mechanics and statistical physics in the bachelor program itself. It is clear that the content of the last three volumes cannot be part of an *integrated course* but rather the subject matter of pure theory lectures. This holds in particular for *Many-Body Theory* which is offered, sometimes under different names, e.g., *Advanced Quantum Mechanics*, in the eighth or so semester of study. In this part, new methods and concepts beyond basic studies are introduced and discussed which are developed in particular for correlated many particle systems which in the meantime have become indispensable for a student pursuing a master's or a higher degree and for being able to read current research literature.

In all the volumes of the series *Theoretical Physics*, numerous exercises are included to deepen the understanding and to help correctly apply the abstractly acquired knowledge. It is obligatory for a student to attempt on his own to adapt and apply the abstract concepts of theoretical physics to solve realistic problems. Detailed solutions to the exercises are given at the end of each volume. The idea is to help a student to overcome any difficulty at a particular step of the solution or to check one's own effort. Importantly these solutions should not seduce the student to follow the *easy way out* as a substitute for his own effort. At the end of each bigger chapter, I have added self-examination questions which shall serve as a self-test and may be useful while preparing for examinations.

I should not forget to thank all the people who have contributed one way or another to the success of the book series. The single volumes arose mainly from lectures which I gave at the universities of Muenster, Wuerzburg, Osnabrueck, and Berlin (Germany), Valladolid (Spain), and Warangal (India). The interest and constructive criticism of the students provided me the decisive motivation for preparing the rather extensive manuscripts. After the publication of the German version, I received a lot of suggestions from numerous colleagues for improvement, and this helped to further develop and enhance the concept and the performance of the series. In particular, I appreciate very much the support by Prof. Dr. A. Ramakanth, a long-standing scientific partner and friend, who helped me in many respects, e.g., what concerns the checking of the translation of the German text into the present English version.

Special thanks are due to the Springer company, in particular to Dr. Th. Schneider and his team. I remember many useful motivations and stimulations. I have the feeling that my books are well taken care of.

Berlin, Germany   Wolfgang Nolting
August 2016

Preface to Volume 5

The main goal of the present volume 5 (*Thermodynamics*) is exactly the same as that of the total course on *Theoretical Physics*. It is thought to be accompanying textbook material for the study of university-level physics. It is aimed to impart, in a compact form, the most important skills of theoretical physics which can be used as basis for handling more sophisticated topics and problems in the advanced study of physics as well as in the subsequent physics research. It is presented in such a way that it enables self-study without the need for a demanding and laborious reference to secondary literature. For the understanding of the text, it is only presumed that the reader has a good grasp of what has been elaborated in the preceding volumes 1, 2, 3, and 4. Mathematical interludes are always presented in a compact and functional form and practiced when they appear indispensable for the further development of the theory. Such mathematical insertions, though, are becoming of course decreasingly necessary with the increasing volume number. For the whole text, it holds that I had to focus on the essentials, presenting them in a detailed and elaborate form, sometimes consciously sacrificing certain elegance. It goes without saying that, after the basic course, secondary literature is needed to deepen the understanding of physics and mathematics. *Thermodynamics* belongs to the classical theories but would thematically be better off as a prelude to *Statistical Mechanics*. The latter can be offered, however, as *modern, nonclassical theory (Quantum Statistics)* only at a later stage of the study, namely, after we have dealt with the *Quantum Mechanics* (volumes 6 and 7). The classical phenomenological *Thermodynamics* takes its concept formation directly from the experiment and does therefore not need, in contrast to the *Quantum Statistics*, any quantum-mechanical element. As a rule, it is a module of the bachelor program in physics and has therefore to be integrated into the first (classical) part of this course on theoretical physics. The exact position of *Thermodynamics* in such a course is, however, not unique. It can also be offered before the *Electrodynamics*. *Thermodynamics* is, as a *science of heat*, a classical phenomenological theory and for the understanding of which, physical terms like *temperature* and *heat* have to be introduced. These quantities are reasonably definable only for macroscopic many-particle systems being completely meaningless for a single particle. The full theory of *Thermodynamics* is based on a few fundamental *theorems or laws*, which have to be considered here as theoretically non-provable but experimentally unrefuted empirical facts. As to these theorems, as well as to the terms temperature and heat, we have to content ourselves, in the framework of thermodynamics, to a certain degree, with an *intuitive self-understanding*. A systematic reasoning is possible only with *Statistical Mechanics* (volume 8) which is thus to be considered as complementary to thermodynamics. It is consistent with, at least in its version as *Quantum Statistics*, the principles of quantum mechanics which will be developed in the volumes 6 and 7. This volume on *Thermodynamics* arose from lectures I gave at the German universities in Muenster, Wuerzburg, and Berlin. The animating interest of the students in my lecture notes has induced me to prepare the text with special care. The present one as well as the other volumes are thought to be the textbook material for the study of basic physics, primarily intended for the students rather than for the teachers. I am thankful to the Springer company, especially to Dr. Th. Schneider, for accepting and supporting the concept of my proposal. The collaboration was always delightful and very professional. A decisive contribution to the book was provided by Prof. Dr. A. Ramakanth from the Kakatiya University of Warangal (India). He deserves a lot of thanks!

Berlin, Germany   Wolfgang Nolting
September 2016

Contents

Chapter 1
Basic Concepts

Thermodynamics is a classical, phenomenological theory (**'science of heat'**) which, as such a theory is expected to, takes its terms and concepts directly from the experiment. It deals with phenomena, to the characterization of which, the physical quantities

temperature and heat

must be referred to. One does not find these quantities either in classical mechanics or in quantum mechanics. They are reasonably definable only for systems which consist of many *'sub-units'*, but are, in contrast, completely meaningless for single objects, as, for instance, the mass point of classical mechanics. Thermodynamics deals exclusively and typically with the study of **macroscopic systems**.

The situation was different in classical mechanics and electrodynamics. There we have discussed and evaluated the fundamental Newton's axioms and the Maxwell equations, respectively, at first on especially simple model systems (mass point, point charge) in order to extend them only subsequently to macroscopic, say, realistic objects. Thermodynamics, in contrast, is conceptualized from the beginning only for macroscopic *many-particle systems*, where it exploits the perplexing fact that such systems, in spite of their many degrees of freedom, are phenomenologically sufficiently well described by only a very few observables such as pressure, volume, temperature, ...

The full theory is based on certain **main theorems**, the so-called **laws of thermodynamics**, out of which all the other statements can be derived. The

zeroth law of thermodynamics

postulates the existence of a **'temperature'**. The

first law of thermodynamics

declares **'heat'** to be a form of energy and requires, under its inclusion, the validity of the energy conservation law. The

second law of thermodynamics

© Springer International Publishing AG 2017
W. Nolting, *Theoretical Physics 5*, DOI 10.1007/978-3-319-47910-1_1

deals with the impossibility to convert heat **completely** into other types of energy as, for instance, mechanical energy of motion. The

third law of thermodynamics

concerns with the unattainability of the absolute zero.

Main statements follow, as we will see, mainly from the first and the second law.

The theory, complimentary to thermodynamics, is the **Statistical Mechanics** which is offered in Vol. 8 of this course on **Theoretical Physics**. This theory rationalizes the terms and laws of the phenomenological thermodynamics via the microscopic structure of the systems, and that using of the concepts of classical mechanics and quantum mechanics, respectively. The principal problems in accomplishing that are obvious: To describe macroscopic systems *microscopically correctly* would mean solving 10^{23} coupled equations of motion with complicated interaction terms. This is impossible, but fortunately also unnecessary since a measuring process always means an averaging (see (2.179), Vol. 3). The decisive challenge of Statistical Mechanics is therefore to fix the relevant macroscopic observables with statistical methods (frequency distributions, average values, probabilities, ...) from the given microscopic data.—A further important task of Statistical Mechanics is to explain basic terms like *temperature* and *heat*, which are, as mentioned, typical for thermodynamics and are directly correlated to the *large number of particles*.

However, this volume deals exclusively with the phenomenological thermodynamics. We therefore have to accept, at first, that the deep, more or less *microscopic* understanding of certain important terms and concepts must be left to a later point of time.

1.1   Thermodynamic Systems

We denote as *thermodynamic system* any macroscopic system which is built up of many *elementary components* (atoms, electrons, photons, field modes, ...). Thermodynamic systems are therefore systems with many degrees of freedom, where one is not interested in any particular microstates.

Examples

A liter of air in the lecture hall,
a galvanic element,
a steam-engine,
a piece of ferromagnetic iron,
a hollow space filled with radiation,
a 'box with contents',
. . .

An important aspect of the thermodynamic systems consists in the possibility of

enclosing them by **walls** *to separate from surroundings,*

so that interactions between system and surroundings can totally or partially be excluded.

In order to understand this aspect better, we must apply in the following, some terms which are quite familiar from the everyday use of language, but can, strictly speaking, be concisely defined for thermodynamics only in subsequent sections:

(A) Isolated system

No exchange whatsoever of 'properties' and 'contents' with the surroundings, i.e., no particle exchange or energy exchange, no interaction with external fields etc.

(B) Closed system

No matter (particle) exchange with the surroundings
Such a system can definitely be still in certain contact with the surroundings.

Examples

1) Heat-exchange contact (thermal contact)

This contact leads to a temperature equalization (thermal balancing) between the system and its surroundings by exchange of energy in the form of heat. If the surroundings can be considered as a *very large* system, whose temperature does not practically change when a finite *'amount of heat'* is taken out or pumped in, then one says that the ('small') system finds itself in contact with a

heat bath.

A system without the contact 1) is called **thermally isolated.**

2) Work-exchange contact

When the system carries out work on the surroundings, or vice versa, certain properties of the system will change. That work can be of mechanical (Fig. 1.1), electromagnetic, chemical or of any other nature.

(C) Open system

There are no restrictions whatsoever, i.e., even particle (matter) exchange is possible.

Fig. 1.1 Compression of a gas by a piston movement as an example of a work-exchange contact

1.2   State, Equilibrium

For the description of a thermodynamic system we use the results of representative measurements on characteristic macroscopic observables, the so-called

state variables

Which quantities eventually come into question is not uniquely predetermined but is rather fixed by our interest and expedience. Criteria for a proper choice may be:

- **simple** measurements,
- **independent** observables,
- sufficiently detailed (**complete**) description, ...

One speaks of a *complete set* of independent state variables if all the other thermodynamic quantities of the system can be represented as functions of these variables. It is a typical feature of thermodynamics that, in spite of many degrees of freedom, only a few state variables already suffice for the description since the atomic (microscopic) composition of the regarded system is immaterial.

Examples

gas-liquid:
  pressure p, volume V, temperature T, particle number N, entropy S, internal energy U ...,
magnet:
  magnetic field $\mathbf{H}$, magnetic moment $\mathbf{m}$, magnetization $\mathbf{M(r)}$, temperature T, ...

Not all the state variables are independent of each other. There exist relations between them. A distinction is therefore drawn between dependent and independent state variables. The dependent ones are called

state functions

One distinguishes:

1) extensive state variables   (*variables of quantity*)
  These are **in proportion to quantity**, i.e., they behave additively when systems are brought together. Examples are:

$$V, \mathbf{m}, \text{ mass } M, U, \ldots$$

2) intensive state variables   (*variables of quality*)
  These are **independent of quantity**, as, for instance, T, p, $\mathbf{M}$, $\rho = N/V$, ...

In thermodynamics one has to deal almost exclusively with either extensive or intensive state variables. Let us now list further important terms:

State space:

Space which is spanned by a **complete set** of independent state variables.

State:

Point in the state space given by the values of a **complete set** of independent state variables.

Equilibrium:

The state for which the values of the basic state variables do not change with time.

According to experience, every isolated system goes *by itself* into a state which does not change in the course of time. This is then the equilibrium state. The time which the system needs to reach this state is called the **relaxation time**. This characteristic time can vary from system to system by orders of magnitude. In classical thermodynamics one always understands by a state an equilibrium state, unless explicitly indicated otherwise.

Change of state, process:

Sequence of states through which the system runs. If the initial state is an equilibrium state, then, a change of state can be induced only by a variation of external parameters.

The change of state is called

quasi-static

if the process is so slow compared to the relevant relaxation times that it practically consists of a sequence of equilibrium states. It describes a curve in the state space.

The change of state is called

reversible

if it is about a reversible sequence of equilibrium states, i.e., if a chronological reversal of the variation of the external conditions leads to a temporal reversal of the equilibrium states which are run through by the system.

Logically, an **irreversible** change of state is not reversible in the above sense. Prime example is the intermixing of two gases (Fig. 1.2). The reintroduction of the dividing wall into the system after the intermixing does not result in demixing and reversal to the initial state. Real processes are normally neither quasi-static nor reversible.

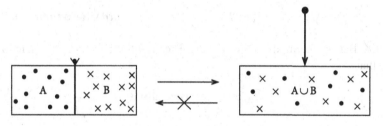

Fig. 1.2 Intermixing of two gases as an example of an irreversible change of state

In the following an important role is played by the

thermodynamic cycle

for which initial and final state are identical, i.e., **all** state variables, not just the independent ones, return to their initial values.

This volume of the basic course in **Theoretical Physics** deals exclusively with equilibrium-thermodynamics (actually better: thermo**statics**). The non-equilibrium thermodynamics is rather involved.

1.3   The Concept of Temperature

Now, when we start to discuss thermodynamic states and thermodynamic processes, we can adopt the meaning and the methods of measuring for most of the observables from other branches of physics as, e.g., mechanics ($p, V, \rho, \ldots$) or electrodynamics ($\mathbf{H}, \mathbf{M}, \ldots$). The concept and the method of measurement temperature, however, we have to introduce completely new. This shall be done in several steps, gradually becoming more precisely.

The term of temperature is of course in a certain sense familiar to us since our birth, in connection with emotional perceptions of *warm* and *cold*. It is therefore, on the one hand, a rather elementary term. On the other hand, however, those sensations of *warm* and *cold* are highly subjective and therefore not reproducible. It is therefore not at all self-evident that *temperature*, too, can be considered as a **physical measurand**. We have to **postulate** its existence!

Definition 1.3.1 (Zeroth Law of Thermodynamics)

1. Each macroscopic system possesses a

   temperature T.

   It is an **intensive** state variable which in an **isolated** system, i.e., a system left to its own, everywhere assumes the same value, which is the homogeneous equilibrium value.
2. T is characterized by **one** number and therefore is a scalar observable.
3. If two systems A and B are in their respective equilibrium, it can always be said:

$$T_A > T_B \quad \text{or} \quad T_A < T_B \quad \text{or} \quad T_A = T_B \quad \textbf{(ordering axiom)} .$$

4. Let A, B, C be thermodynamic systems. Then it follows from $T_A > T_B$ and $T_B > T_C$ that

$$T_A > T_C \quad \textbf{(transitivity)} .$$

5. Let the systems **A** and **B** be in *thermal contact*, while the total system A ∪ B is isolated, then it holds in the equilibrium:

$$T_A = T_B = T_{A \cup B} \, .$$

6. If two systems are separated and

$$T_A^{(i)} < T_B^{(i)} \, ,$$

then after establishing thermal contact, in the equilibrium:

$$T_A^{(f)} < T_{A \cup B} < T_B^{(f)} \, .$$

As temporary measuring prescription we use the impact of temperature on other observables. Each physical property, which changes monotonically and uniquely with T, can be used to construct a **thermometer**,

> mercury thermometer   (volume) ,
> gas thermometer   (pressure) ,
> resistance thermometer (electric resistance) .

Details on the way they function should be taken from textbooks on experimental physics.

One should bear in mind that each measurement of temperature uses in a very decisive manner the property 5. of the *thermal equilibrium*. Each thermometer measures actually its own temperature which only in the thermal equilibrium agrees with that of the system under study. In case of different initial temperatures, then always because of 6., there appears a certain misrepresentation of the true temperature of the system.

1.4   Equations of State

By *equations of state* we understand relations between certain extensive and intensive state variables Z_i of the system:

$$f(Z_1, Z_2, \dots, Z_n) = 0 \, . \tag{1.1}$$

They must be uniquely reversible, i.e., solvable for all variables Z_i. Using them one can convert dependent into independent state variables and vice versa.

The equations of state applied in thermodynamics are accepted without derivation as experimentally verified matters of fact. They follow generally from simple theoretical models which mimic the underlying physical systems. We want to briefly broach here the four most important examples.

1.4.1   Ideal Gas

We start with the simplest system, a gas of N molecules, that fulfills the following two strongly idealized **assumptions**:

1. the molecules have no volume of their own (mass points),
2. the particles do not interact with one another.

These presumptions are of course fulfilled in a real gas, strictly speaking, only in the case of infinite dilution.

Let a body of gas with N particles be enclosed in a volume V. It may be in thermal contact with a heat bath of a definite temperature. According to the zeroth law the gas takes on, in the equilibrium, the same temperature as the bath. In the gas there is a homogeneous pressure p. With a change of the volume V the pressure p will also change.

Experimental Observation

In the case of sufficient dilution $\rho = N/V \to 0$ all gases behave in the same manner and obey the

Boyle-Mariotte's Law

$$\frac{pV}{N} = K = \text{const} . \tag{1.2}$$

One can interpret (1.2) as the defining equation for the **ideal gas**. The constant K assumes different values for heat baths of different temperatures and can therefore serve to fix a measuring prescription for the temperature.

Ansatz

$$K(\vartheta) = K_0(1 + \alpha\,\vartheta) . \tag{1.3}$$

Celsius (Centigrade) Scale

$$\vartheta = 0\,°\text{C} : \quad \text{freezing point of water} ,$$
$$\vartheta = 100\,°\text{C} : \quad \text{boiling point of water at } p = 1\,\text{atm} .$$

It follows with the measuring values for $K_0 = K(0\,°C)$ and $K(100\,°C)$:

$$\alpha = \frac{K(100\,°C) - K(0\,°C)}{100\,°C\,K(0°C)} = \frac{1}{273.2} . \tag{1.4}$$

This result is independent of the type of the gas as long as (1.2) is valid. By use of (1.2) to (1.4) the temperature of each heat bath and gas, respectively, can be determined.

Kelvin Scale (Absolute Temperature)

$$T = \alpha^{-1} + \vartheta = 273.2\,K + \vartheta . \tag{1.5}$$

The constant

$$k_B = K_0\,\alpha$$

is universal (**Boltzmann constant**) and has the value

$$k_B = 1.3805 \cdot 10^{-23}\,J/K . \tag{1.6}$$

Therewith we have the
Equation of state of the ideal gas

$$p V = N k_B T . \tag{1.7}$$

This equation can further be reformulated a bit if one uses the **Avogadro constant** (formerly: Loschmidt number)

$$N_A = 6.02252 \cdot 10^{23}\,mol^{-1} , \tag{1.8}$$

or the **universal gas constant**:

$$R = k_B N_A = 8.3166 \frac{J}{mol\,K} , \tag{1.9}$$

If one denotes by $n = N/N_A$ the number of moles then (1.7) can also be written as:

$$p V = n R T . \tag{1.10}$$

The so defined temperature scale has a universal character since (1.2) is independent of the type of the ideal gas. p, V, N are positive quantities and therewith the temperature T, too. A disadvantage of this definition of temperature lies of course in the fact that it is bound to gases which fulfill the ideal-gas equation. It thus needs the two conditions 1. and 2. The definition becomes certainly useless for $T \to 0$ and/or large p because of the then appearing liquefaction. This drawback will be removed by the introduction of a universal (thermodynamic) temperature that succeeds in connection with the second law of thermodynamics. The concept of temperature formulated in this section will therefore be only of provisional character.

1.4.2   Van der Waals Gas

The equation of state of the ideal gas (1.7) can be applicable for real gases, because of the two restrictions 1. and 2., only in the limit of very low particle densities. In particular, it fails to describe the phase transition *'gas↔ liquid'*. By the following **ansatz**

$$p_{\text{eff}} V_{\text{eff}} = n R T \tag{1.11}$$

we want the generalize the ideal-gas equation (1.10) in such a way that, on the one hand, the two restrictions 1. and 2. are removed, but, on the other, it is also such that in the limit of very strong dilution (1.10) is reproduced.

To 1

For $p \to \infty$ and $T = $ const the ideal-gas equation predicts $V \to 0$ therewith disregarding the fact that the gas molecules have a *'volume of their own'* (*proper volume*). For the real gas we will have to take into consideration a minimal volume:

$$V_{\text{min}} \approx N \times \text{'particle volume'} \equiv \frac{b}{N_{\text{A}}} N$$

$$\implies \quad V_{\text{eff}} = V - V_{\text{min}} = V - n b . \tag{1.12}$$

We denote the correction term nb as **'proper volume'**.

To 2

The particles of the real gas, confined to a finite box, interact with each other. Because of the homogeneous distribution the interaction forces of the particles cancel each other out in the inside space of the box. For a particle at the edge of the box, though, there remains a resulting force component inwards (Fig. 1.3). That diminishes the pressure of the gas on the walls of the box where, otherwise, the pressure is measured:

$$p_{\text{eff}} > \text{'wall-pressure'} \; p .$$

Fig. 1.3 The justification of the internal pressure in the van der Waals-equation of state

The difference pressure is proportional to the number of particle interactions in the wall layer, whose thickness d is approximately given by the average range of these interactions. The number of particle interactions in the edge layer can be estimated by:

$$N'\left(N'-1\right) \sim \left(d \cdot S\right)^2 \left(\frac{N}{V}\right)^2 \sim \left(\frac{N}{V}\right)^2 \qquad (S : \text{surface of the box}).$$

This corresponds to

$$p_{\text{eff}} = p + a\,\frac{n^2}{V^2}\,.\tag{1.13}$$

We denote the correction term an^2/V^2 as the **'intrinsic pressure'**. If we insert the equations (1.12) and (1.13) into (1.11) we get the **van der Waals equation of state**

$$\left(p + a\,\frac{n^2}{V^2}\right)(V - nb) = nRT\,.\tag{1.14}$$

a and b are phenomenological material constants, where a varies very strongly, b less strongly from substance to substance.—Let us inspect the equation of state (1.14) in a bit more detail:

1) Critical point

One can cast (1.14) into the following form:

$$V^3 - V^2\left(nb + \frac{nRT}{p}\right) + V\,\frac{an^2}{p} - ab\,\frac{n^3}{p} = 0\,.\tag{1.15}$$

This is an equation of third degree for the volume V, which for given p, T such that $p < p_c$, $T < T_c$ has three real solutions, and otherwise, one real and two complex solutions. Consequently, there exists a **critical point**

$$(p_c, V_c, T_c)\,,$$

at which the three solutions just coincide. At this special point it must then hold:

$$0 \stackrel{!}{=} (V - V_c)^3 = V^3 - 3V^2 V_c + 3V\,V_c^2 - V_c^3\,.$$

Comparison of coefficients with (1.15) leads to the critical data of the real gas, which are all determined by the two phenomenological parameters a, b:

$$V_c = 3bn; \quad p_c = \frac{a}{27b^2}; \quad RT_c = \frac{8a}{27b}. \tag{1.16}$$

One can of course eliminate a and b from these equations to get:

$$Z_c = \frac{p_c V_c}{nRT_c} = \frac{3}{8}. \tag{1.17}$$

Experimentally one finds that for almost all real gases $Z_c < 3/8$, while for the ideal gas $Z_c = 1$ (1.10). On that score, the van der Waals model represents an obvious improvement.

2) The law of corresponding states

If one introduces the *reduced* quantities

$$\pi = \frac{p}{p_c}; \quad v = \frac{V}{V_c}; \quad t = \frac{T}{T_c} \tag{1.18}$$

then one can bring the van der Waals equation (1.14) into a form which no longer contains any material constant and therefore should be valid for **all** substances (see Exercise 1.6.6):

$$\left(\pi + \frac{3}{v^2}\right)(3v - 1) = 8t. \tag{1.19}$$

One says that two substances with the same (π, v, t)-values are in **corresponding states**. This universal equation is in general far better fulfilled than the original van der Waals equation (1.14), from which it was derived.

3) Maxwell construction

The pV-isotherms show for $T < T_c$ an unphysical peculiarity. There exists a region where (Fig. 1.4)

$$\left(\frac{\partial p}{\partial V}\right)_T > 0.$$

This cannot be realistic since a volume decrease $dV < 0$ would then bring about also a pressure decrease $dp < 0$. The system would collapse. The reason for this unphysical feature can be traced back to the fact that we implicitly assumed, when deriving the van der Waals equation of state, that the system consists of exactly **one** homogeneous phase. We denote a phase as **homogeneous** if the intensive state variables, such as for instance $\rho, T, p, \ldots,$, have the same numerical values everywhere in it. This assumption turns out to be not correct for $T < T_c$. In the

Fig. 1.4 Isotherms of the van der Waals gas

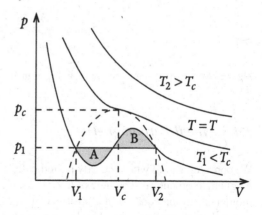

shaded region in Fig. 1.4, in fact, a two-phase region is present. Liquid and gas coexist in equilibrium. Later we will learn that the van der Waals isotherm here is to be replaced by a parallel to the V-axis, and that too in such a way that the areas A and B in Fig. 1.4 are equal. One calls this the **Maxwell construction**, the physical justification for this will be provided at a later stage.

At the temperature T_1, the pressure p_1 and the volume V_1 in Fig. 1.4, the system consists of one homogeneous phase of liquid. If then at this fixed temperature the volume is increased, the pressure remains constant, and a part of the liquid vaporizes into gas. At V_2 the whole system is gaseous. A further volume increase then leads to a pressure decrease.

4) Virial expansion

Phase transitions are obviously of discontinuous nature as we have just seen in connection with the transition liquid $\Longleftrightarrow$ gas. It is therefore not to be expected that **exact** equations of state of real gases are represented by simple analytical expressions. Even for a first approximation description in the form of the van der Waals equation, we had to bring in the Maxwell construction. One therefore sometimes uses series expansions in the particle density:

$$p = \frac{N k_{\mathrm{B}} T}{V} \left\{ 1 + B_1 \left(\frac{N}{V} \right) + B_2 \left(\frac{N}{V} \right)^2 + \dots \right\} . \tag{1.20}$$

The so-called **virial coefficients** B_i express the deviation from the behavior of the ideal gas. They are theoretically justified in the framework of Statistical Mechanics. For the *van der Waals gas* one finds:

$$B_1 = \frac{b}{N_{\mathrm{A}}} - \frac{a}{N_{\mathrm{A}}^2 k_{\mathrm{B}} T} ; \quad B_\nu = \left(\frac{b}{N_{\mathrm{A}}} \right)^\nu , \quad \text{for } \nu \geq 2 . \tag{1.21}$$

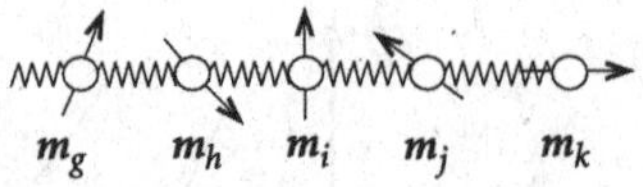

$$m_g \qquad m_h \quad m_i \quad m_j \qquad m_k$$

Fig. 1.5 Model of a paramagnet, built up by magnetic moments $\mathbf{m}_i$ localized at certain 'lattice sites'

1.4.3 Ideal Paramagnet

We consider a solid whose strictly periodically ordered ions each carry a permanent magnetic moment $\mathbf{m}_i$. The index i ($i = 1, 2, \ldots, N$) numbers serially the individual moments. An ion consists of a positively charged nucleus around which negatively charged electrons circulate. The latter represent micro-circular currents with each of which is associated, according to ((3.43), Vol. 3), a resulting magnetic moment. Since the moments are vectors and are oriented along random directions, in most cases they compensate each other in their actions. In some solids, however, there remains a net magnetic moment per ion which is indicated in Fig. 1.5 by a vector arrow. A possible interaction between the localized moments is symbolized by small springs.—One can find some physical details about magnetic solids in section 3.4, Vol. 3.—In thermodynamics, though, we will use only very simple models of paramagnetism and ferromagnetism.

In the case of an *ideal paramagnet* one starts, as for the ideal gas, with the assumption that there is **no interaction** between the moments. The directions of the moments are statistically distributed, so that normally the resulting total moment vanishes. However, if one switches on a homogeneous magnetic field,

$$\mathbf{H} = H\,\mathbf{e}_z \, ,$$

the elemental dipoles try to align themselves parallel to the field because of ((3.52), Vol. 3). As a consequence of this 'alignment-effect' there arises a finite macroscopic total moment, which, however, exhibits a temperature-dependence. Later we will understand why the thermal energy of the elemental magnets, increasing with T, is opposed to the 'alignment-effect'. We find the equation of state of the ideal paramagnet by use of the definitions:

total moment : $\mathbf{m}_{\text{tot}} = \sum_i \mathbf{m}_i$; $|\mathbf{m}_i| = m$ $\forall i$,
magnetization : $\mathbf{M} = \frac{1}{V}\mathbf{m}_{\text{tot}} = M(T, \mathbf{H})\,\mathbf{e}_z$.

Since $\mathbf{H}$ and $\mathbf{M}$ are in general parallel, we can suppress the vector-notation. The **equation of state of the ideal paramagnet** then reads:

$$M = M_0\, L\left(m\frac{B_0}{k_{\text{B}} T}\right) , \tag{1.22}$$

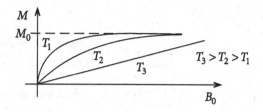

Fig. 1.6 Magnetization of a paramagnet as function of an external magnetic induction for three different temperatures. M_0: saturation magnetization

$$L(x) = \coth x - \frac{1}{x} \; : \; \textbf{Langevin function} , \qquad (1.23)$$

$$M_0 = \frac{N}{V}m \; : \; \textbf{saturation magnetization} , \qquad (1.24)$$

$$B_0 = \mu_0 H .$$

Equation (1.22) can easily be derived by the means of Statistical Mechanics. We content ourselves here, however, with the result.

The *saturation* is reached when all the moments are oriented parallel (Fig. 1.6). Then a further enhancement of the field does not lead to an increase of the magnetization. It is typical for a paramagnet:

$$M(H,T) \xrightarrow[H \to 0]{} 0 .$$

For high temperatures the argument of the Langevin function becomes very small. Then we can simplify further on, because of $L(x) \xrightarrow[x \to 0]{} x/3$ (1.22):

$$M = \frac{C}{T} H \qquad \textbf{Curie law} . \qquad (1.25)$$

C is the so-called *Curie constant*,

$$C = \mu_0 \frac{N}{V} \frac{m^2}{3 k_B} . \qquad (1.26)$$

Normally one uses the equation of state of the ideal paramagnet in the simplified form (1.25).

1.4.4 Weiss Ferromagnet

The ferromagnet is the *magnetic analog* of the real gas. A so-called **exchange interaction** takes care for the existence of a critical temperature denoted as *Curie* *temperature*:

$\boxed{T \le T_c}$:   *spontaneous* magnetization
$$M(T, H) \xrightarrow[H \to 0]{} M_S(T) \ne 0 ,$$

$\boxed{T > T_C}$:   properties as those of the paramagnet.

The exchange interaction, which is so far not completely understood, can be simulated to a good approximation, insofar as its impacts are concerned, by an effective magnetic field,

$$B_{\text{eff}} = \mu_0 \, \lambda \, M , \tag{1.27}$$

which is assumed to be proportional to the magnetization. λ is the so-called **exchange constant**. One thus replaces the interacting moments by non-interacting moments in an effective field of the form B_{eff}. This adds to the external field B_0. Except for this, we have the same situation as for the paramagnet which allows us to use (1.22) in a properly modified version:

$$M\,(T, B_0) = M_0 \, L \left[m \frac{B_0 + B_{\text{eff}}}{k_B \, T} \right] . \tag{1.28}$$

This is the **equation of state of the ferromagnet**, which represents an implicit conditional equation for the magnetization.

The **spontaneous** magnetization $M_S(T)$ follows from (1.28) for $B_0 = 0$:

$$M_S(T) = M_0 \, L \left[\mu_0 \, m \, \frac{\lambda \, M_S(T)}{k_B \, T} \right] . \tag{1.29}$$

One recognizes that $M_S(T) \equiv 0$ (paramagnetism!) is always a solution. Under certain circumstances, however, there are still further solutions. Since $L(x)$ in (1.29) saturates as function of M for large M at 1, a solution $M_S \ne 0$ appears exactly then when the right-hand side of (1.29) has an initial slope greater than 1 (Fig. 1.7):

$$1 \overset{!}{\le} \frac{d}{dM} \left(M_0 \, L \left[\mu_0 \, m \, \frac{\lambda \, M(T)}{k_B \, T} \right] \right) \Bigg|_{M = 0}$$

$$= \frac{d}{dM} \left(M_0 \mu_0 m \, \frac{\lambda \, M}{3 \, k_B \, T} \right) \Bigg|_{M = 0} = \frac{N}{V} \mu_0 m^2 \frac{\lambda}{3 k_B \, T} = \frac{\lambda \, C}{T} .$$

Fig. 1.7 Graphical determination of the spontaneous magnetization of the Weiss ferromagnet

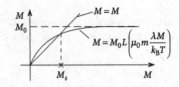

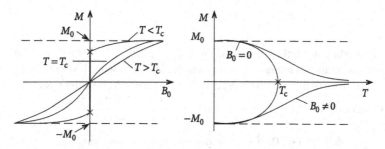

Fig. 1.8 '*Left*': Field dependence of the magnetization of the ferromagnet for temperatures above, below and equal to the Curie temperature.   '*Right*': Temperature dependence of the magnetization with and without external field

Therewith, the Curie temperature is fixed:

$$T_c = \lambda C . \qquad (1.30)$$

$$\boxed{T < T_c} \iff \tfrac{\lambda C}{T} > 1 \iff M_S \neq 0 \text{ does exist}$$
$$\iff \textbf{ferromagnetism} ,$$
$$\boxed{T > T_c} \iff \tfrac{\lambda C}{T} < 1 \iff \text{only } M_S = 0 \text{is a solution}$$
$$\iff \textbf{paramagnetism} .$$

If there exists a solution $M_S > 0$, then $-M_S$ is also a solution, which is, because of $\tanh(-x) = -\tanh x$, easily realized with (1.29).

The more exact analysis of (1.28) yields qualitatively the behavior exhibited in Fig. 1.8. The reader should take further details on paramagnetism and ferromagnetism from the respective special literature.

Just as for the paramagnet, one can still further simplify the equation of state (1.28) if one restricts the considerations to *high temperatures* and *low fields*:

$$M \approx M_0\, m\, \frac{B_0 + B_{\text{eff}}}{3\, k_{\text{B}}\, T} = \frac{C}{T}(H + \lambda M) = \frac{C}{T}H + \frac{T_c}{T}M .$$

From this it follows the
Curie-Weiss law

$$M(H, T) \approx \frac{C}{T - T_C} H . \qquad (1.31)$$

For the discussion of thermodynamics we will content ourselves with this simplified version of the equation of state of a ferromagnet.

While discussing magnetic systems, in general, pressure and volume effects are not taken into consideration.

1.5 Work

Changes of state are linked in general to energy changes. In thermodynamics one typically distinguishes energy changes, which are due to work (ΔW) done by or on the system, from those, which change the heat content (ΔQ) of the system.

Sign Convention

$$\Delta W > (<) \; 0, \text{ if work is done on (by) the system },$$
$$\Delta Q > (<) \; 0, \text{ if heat is pumped into (out of) the system }.$$

The term 'work' is taken from Classical Mechanics or Electrodynamics, depending on the context:

$$q_1, \ldots, q_m : \quad \text{generalized coordinates },$$
$$F_1, \ldots, F_m : \quad \text{(associated) generalized force components }.$$

The q_i need not necessarily have the dimension *'length'*, and the *'forces'* F_i need not necessarily be forces in the strict sense. However, the product $q\,F$ has to have the dimension of energy.

(Differential, Quasi-Static) Work

$$\delta W = \sum_{i=1}^{m} F_i \, dq_i . \tag{1.32}$$

Examples

	F_i	dq_i	δW
pressure	$-p$	dV	$-p\,dV$
surface tension	σ	dS	$\sigma\,dS$
magnetic field	$\mathbf{B}_0$	$d\mathbf{m}$	$\mathbf{B}_0 \cdot d\mathbf{m}$
electric field	$\mathbf{E}$	$d\mathbf{P}$	$\mathbf{E} \cdot d\mathbf{P}$
chemical potential	μ	dN	$\mu\,dN$

(V: volume, S: surface, $\mathbf{m}$: magnetic moment, $\mathbf{P}$: electric polarization, N: particle number)

For the differential work we have consciously chosen the Greek letter 'δ' instead of the usual '**d**' in order to indicate that δW does **not always** represent a total differential. That means that line integrals $\int_C \delta W$ are in general path-dependent.

Incidental remark:

A differential form

$$\delta A = \sum_{j=1}^{m} a_j (x_1, x_2, \ldots, x_m) \, dx_j$$

is a **total differential (integrable)** if and only if the following **integrability conditions** are fulfilled (cf. (2.235), Vol. 1):

$$\left(\frac{\partial a_j}{\partial x_i} \right)_{x_m, m \neq i} = \left(\frac{\partial a_i}{\partial x_j} \right)_{x_m, m \neq j} \qquad \forall i, j \, . \tag{1.33}$$

Then it holds for each closed path (see Exercise 1.6.3):

$$\oint \delta A = \oint dA = 0 \, .$$

Let $m = 2$, as is frequently the case in thermodynamics, and let

$$\delta A = a_1 \, dx_1 + a_2 \, dx_2$$

be **not** a total differential, then there does **always** exist an **'integrating factor'** $\mu(x_1, x_2)$, such that

$$df = \mu \, \delta A = (\mu \, a_1) \, dx_1 + (\mu \, a_2) \, dx_2$$

becomes a total differential. For this purpose μ must be fixed such that

$$\left(\frac{\partial (\mu \, a_1)}{\partial x_2} \right)_{x_1} = \left(\frac{\partial (\mu \, a_2)}{\partial x_1} \right)_{x_2} \, . \tag{1.34}$$

The choice of $\mu = \mu(x_1, x_2)$ is not unique (cf. section 2.4.2, Vol. 1.)

The state variables (state functions) of thermodynamics must be unique. If the system passes through a closed path in the state space then all the dependent as well as independent state variables must come back to their initial values. From a state variable η we thus require that

$$\oint d\eta = 0$$

for all closed paths in the state space. That means that $d\eta$ must be a total differential.

The *work*, which is a form of energy, is in this sense **not** a state quantity. The generalized forces F_i in general depend also on the state variable *temperature*. Then

Fig. 1.9 To the term of
volume work on a gas

that is true also for δW, so that we should actually write instead of (1.32):

$$\delta W = \sum_{i=1}^{m} F_i(q_1, \ldots, q_m, T)\, dq_i + 0\, dT\ .$$

The integrability conditions (1.33) cannot then be satisfied:

$$\left(\frac{\partial Q_i}{\partial T}\right)_{\ldots} \neq \left(\frac{\partial 0}{\partial q_i}\right)_{\ldots} = 0\ .$$

Let us inspect in more detail the most important types of *'work'*.

1) Volume work

A gas may be in a cylindrical vessel with the cross section F. The vessel is bounded above by a friction-less moving piston, on which a weight of mass M is located (Fig. 1.9). Equilibrium is given as soon as the gas pressure p can balance the weight:

$$pF = Mg\ .$$

By an infinitesimal shift upwards of the weight its potential energy increases by

$$E_{\text{pot}} = M g\, dx = p F\, dx = p\, dV\ .$$

The work carried out by the gas onto the weight, i.e. the work done outwards, is thus:

$$\delta W = -p\, dV\ . \tag{1.35}$$

For a finite change of the volume $V_1 \to V_2 > V_1$ we then have:

$$\Delta W_{12} = -\int_{V_1}^{V_2} p(V)\, dV\ . \tag{1.36}$$

How would one perform a corresponding experiment? The pressure p is measured via the mass M on the piston. However, in order that the gas can actually expand ($V_2 > V_1$), the piston pressure must be a little bit smaller than the gas pressure. In the experiment the system moves therefore along the line (a) in Fig. 1.10. If the pressure-difference is large, then a rapid expansion of the gas will be the consequence. It gains

Fig. 1.10 Practical realization of the measurement of the volume work referring to the piston pressure in Fig. 1.9

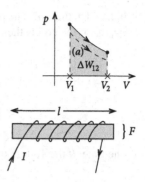

Fig. 1.11 Magnetic system in the inside of a current carrying coil. Illustration of the term 'magnetization work'

flow energy which eventually turns into the still to be discussed *heat energy*. The work done during the change of volume is then smaller than the integral in (1.36). Only if one accepts a sufficiently slow course of the volume change, i.e. performing the experiment **'quasi-statically'**, the pressure-difference can be made arbitrarily small so that (a) coincides with the real $p(V)$-curve (Fig. 1.10). That corresponds then obviously to a maximal performance of work.

2) Magnetization work

We want to justify the expression

$$\delta W = \mathbf{B}_0 \cdot d\mathbf{m} . \tag{1.37}$$

m is the total magnetic moment of the system which is considered here as a state variable. That appears to be not unproblematic since for the creation of a finite moment, for instance in a paramagnet, an external magnetic field **H** is unavoidable, through which, even in the vacuum, a field energy comes into play. However, this energy contribution should **not** be counted in since **H** is no more than an auxiliary means to create the moment **m**. We are focused only on **that** contribution which stems exclusively from the magnetization. This we want to illustrate by an example:

Let the system be in the core of a long, thin, current carrying coil with N turns (Fig. 1.11). **H, B, M, m**, in this case, are all oriented along axial direction, where for the fields we have according to ((4.68),Vol. 3):

$$B = \mu_r \mu_0 \frac{N}{l} I ; \quad H = \frac{N}{l} I .$$

The magnetic flux through the cross-section F amounts to $\Phi = BF$. According to the law of induction the voltage

$$U = -N \dot{\Phi} = -N F \dot{B}$$

is induced in the coil. The power acting on the charges which are transported through the turns of the coil is then:

$$P = UI = -NFI\dot{B} = -lHF\dot{B} = -VH\dot{B}$$
$$= -VH\mu_0 (\dot{H} + \dot{M})$$
$$= -\frac{\mu_0}{2} V \frac{d}{dt} H^2 - \mu_0 VH \frac{d}{dt} M .$$

In the time dt the system thus executes the following work on the charge carriers:

$$- \delta W^* = -\frac{\mu_0}{2} V \, dH^2 - V B_0 \, dM . \tag{1.38}$$

The first term represents the mentioned magnetic field energy which we are not interested in. In a gedanken-experiment we *clamp* the elemental magnets so that the magnetization does not change when we switch off subsequently the auxiliary field H. Thereby we get back the field energy of the vacuum. It remains as **pure magnetization work**:

$$\delta W = \delta W^* - \frac{\mu_0}{2} V \, dH^2 = V B_0 \, dM = B_0 \, dm .$$

That explains (1.37). Notice that here the magnetic induction of the vacuum $B_0 = \mu_0 H$ appears and **not** that of the matter ($B = \mu_r \mu_0 H$)!

1.6   Exercises

Exercise 1.6.1 Investigate whether df represents a total differential:

1. $df = \cos x \sin y \, dx - \sin x \cos y \, dy$.
2. $df = \sin x \cos y \, dx + \cos x \sin y \, dy$.
3. $df = x^3 y^2 dx - y^3 x^2 dy$.

Exercise 1.6.2 Let x, y, z be quantities which obey a functional connection of the form

$$f(x, y, z) = 0 .$$

Verify the following relations:

1. $\left(\dfrac{\partial x}{\partial y}\right)_z = \dfrac{1}{\left(\dfrac{\partial y}{\partial x}\right)_z}$.

2. $\left(\dfrac{\partial x}{\partial y}\right)_z \left(\dfrac{\partial y}{\partial z}\right)_x \left(\dfrac{\partial z}{\partial x}\right)_y = -1$.

Fig. 1.12 Paths between two
fixed points in the xy-plane

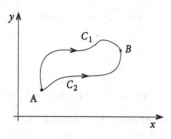

Exercise 1.6.3

1. The path integral

$$I(C) = \int_{A(C)}^{B} \left\{ \alpha(x, y) \, dx + \beta(x, y) \, dy \right\}$$

is to be calculated for fixed end points A and B in the xy-plane. It will have, in general, different values for different paths C_i between A and B (Fig. 1.12). Demonstrate that I is independent of the path if and only if

$$\frac{\partial \alpha}{\partial y} = \frac{\partial \beta}{\partial x} \, .$$

2. Choose especially

$$\alpha(x, y) = y^2 e^x \, ; \quad \beta(x, y) = 2ye^x \, .$$

Calculate

$$I_{AB} = \int_{A}^{B} \left\{ \alpha(x, y) \, dx + \beta(x, y) \, dy \right\}$$

for $A = (0, 0); B = (1, 1)$. Consider at first whether the problem is reasonable at all. If yes, then find the value of I_{AB}!
3. Do the same investigations as under 2., but with interchanged roles of α and β:

$$\alpha(x, y) = 2ye^x \, ; \quad \beta(x, y) = y^2 e^x$$

Exercise 1.6.4 Assuming the equation of state of a gas,

$$p = p(V, T) \, ,$$

is given, express the isobaric, thermal volume expansion coefficient,

$$\beta = \frac{1}{V} \left(\frac{\partial V}{\partial T} \right)_p \, ,$$

and the isothermal compressibility,

$$\kappa_T = -\frac{1}{V}\left(\frac{\partial V}{\partial p}\right)_T ,$$

by the partial derivatives of p with respect to V and T.

Exercise 1.6.5 For a homogeneous substance with the mole number n the following relations have been found:

$$\beta = \frac{nR}{pV} ; \quad \kappa_T = \frac{1}{p} + \frac{a}{V} .$$

β and κ_T are defined as in Exercise 1.6.4, R is the universal gas constant.
    Find the equation of state

$$f(T, p, V) = 0 !$$

Exercise 1.6.6 The van der Waals equation (1.14) describes qualitatively the transition gas $\rightarrow$ liquid.

1. Express the constants a and b in the van der Waals equation by V_c and T_c.
2. Rewrite the van der Waals equation in terms of the *reduced* quantities:

$$\pi = \frac{p}{p_c} ; \quad v = \frac{V}{V_c} ; \quad t = \frac{T}{T_c} .$$

3. Calculate the isothermal compressibility

$$\kappa_T = -\frac{1}{V}\left(\frac{\partial V}{\partial p}\right)_T$$

   for $V = V_c$. Which behavior does κ_T show, when the temperature approaches T_c from above? How can one understand physically this behavior?
4. Investigate, as in 3., the isobaric volume expansion coefficient:

$$\beta = \frac{1}{V}\left(\frac{\partial V}{\partial T}\right)_p .$$

Exercise 1.6.7 The thermal equation of state of a real gas is given by

$$p = nRT(V - nb)^{-1} e^{-\frac{na}{RTV}} .$$

Let $n = N/N_A$ be the number of moles, R the universal gas constant, and a, b material constants (*Dieterici gas*).

1. Calculate from the virial expansion in the particle density $\rho = N/V$,

$$p = k_B T \rho \left(1 + \sum_{\nu=1}^{\infty} B_\nu \, \rho^\nu \right) ,$$

the first coefficient B_1. Express the *Boyle temperature* T_B, for which holds $B_1 = 0$, by the constants a and b.

2. Compare the expression for B_1 with the corresponding virial coefficient of the van der Waals equation. Which meaning do the quantities a and b have?

3. How is the quantity $\left(\frac{\partial p}{\partial \rho} \right)_T$ related to the isothermal compressibility κ_T? Which sign has to be expected, on grounds of physical arguments, for $\left(\frac{\partial p}{\partial \rho} \right)_T$?

4. Calculate $\left(\frac{\partial p}{\partial \rho} \right)_T$ for the Dieterici gas and determine the temperature $T_0(\rho)$, for which this differential quotient becomes zero. Sketch $T_0(\rho)$. Find the *critical temperature* T_c as the maximum of $T_0(\rho)$. Express the quantities a and b by T_c and the *critical* density ρ_c. Which connection exists between T_c and T_B?

5. Plot qualitatively the isotherms, as they come out from the Dieterici equation, in the p-ρ-diagram. In which region are the curves unphysical? For which temperatures can the gas be converted into liquid by increasing pressure?

Exercise 1.6.8   The deviations, seen in the behavior of real gases, from that of an ideal gas are taken into consideration approximately by several types of equations of state:

1. Provision of the proper volume:

$$p(V - nb) = nRT .$$

2. Virial expansion with respect to the pressure:

$$pV = nRT(1 + A_1 p) ; \quad A_1 = A_1(T) .$$

3. Virial expansion with respect to the volume:

$$pV = nRT \left(1 + \frac{B_1}{V} \right) ; \quad B_1 = B_1(T) .$$

Calculate for these three versions the isothermal compressibility κ_T and the isobaric thermal volume expansion β (see Exercise 1.6.4) and compare them with the results for the ideal gas.

Exercise 1.6.9   Calculate the work performed on an ideal paramagnet when the magnetic field H is isothermally ($T = $ const) enhanced from H_1 to $H_2 > H_1$.

Exercise 1.6.10   The plates of a parallel-plate capacitor with the capacity C carry the charges Q and $-Q$. When the plates approach each other a bit then the capacity increases by dC.

1. Which mechanical work must be done thereby?
2. How does the field energy in the capacitor change?
3. Is the change of state reversible?

Exercise 1.6.11   Let

$$B_0 = B_0(T; m) = \alpha T \ln \left(\frac{m_0 + m}{m_0 - m} \right) - \gamma m \ .$$

be a model-equation of state of a ferromagnet. Thereby, $B_0 = \mu_0 H$ is the magnetic induction, m the magnetic moment, $m_0 > 0$ the saturation moment ($-m_0 \leq m \leq +m_0$) and α and γ positive constants.

1. For which values of (T, m) does the ferromagnetic phase become unstable? That is signalled when an increase of the field strength leads to a reduction of the magnetic moment:

$$\left(\frac{\partial B_0}{\partial m} \right)_T (T, m) < 0 \ .$$

   Use the abbreviation:

$$T_C = \frac{\gamma m_0}{2\alpha} \ .$$

2. Calculate the limiting curve $m_S = m_S(T)$, which separates the stable from the unstable region being thus defined by

$$\left(\frac{\partial B_0}{\partial m} \right)_T = 0 \ .$$

3. Show that for $T > T_C$ $m = 0$ is the only solution of $B_0(T, m) = 0$ (paramagnetic phase). Use a graphical construction, which shows that for $T < T_C$ two further zeros $\pm m_S \neq 0$ exist (ferromagnetic phase).
4. Plot qualitatively the isotherms in the B_0, m-diagram!

1.7   Self-Examination Questions

To Sect. 1.1

1. What do we understand by a *thermodynamic system*?
2. When is a system denoted as *isolated, closed* or *open*?
3. Comment on the terms *work-exchange contact, heat-exchange contact* and *heat bath*.

To Sect. 1.2

1. When do we speak of a *complete set* of independent state variables?
2. What do we understand by extensive or intensive state variables?
3. When is a thermodynamic system in its equilibrium?
4. How do we interpret the term *state* in the phenomenological thermodynamics?
5. What is a process? When is it *quasi-static, reversible, irreversible*?
6. What does one understand by a *thermodynamic cycle*?

To Sect. 1.3

1. What follows from the zeroth law of thermodynamics?
2. Characterize the state variable *temperature*.
3. What is to be understood by the transitivity of the temperature?
4. Which property of the temperature is decisive for the mode of operation of a thermometer?

To Sect. 1.4

1. What is an *equation of state*?
2. Which assumptions define the *ideal gas*?
3. How can we fix a measuring prescription for the temperature by use of Boyle-Mariotte's law?
4. How are Celsius and Kelvin scales defined?
5. How does the equation of state of the *ideal gas* read?
6. Which connection exists between the Boltzmann constant and the universal gas constant?
7. By which ansatz does one try to generalize the ideal gas equation to real gases?
8. What do we understand in connection with the real gas by the terms *proper volume* and *intrinsic pressure*?
9. Find justification for the van der Waals equation of state.
10. How can we fix the data p_c, V_c, T_c of the critical point by the van der Waals constants a and b?
11. When are two different real gases in *corresponding states*?
12. Which unphysical result of the van der Waals model is corrected by the *Maxwell construction*?
13. What does one understand by *virial expansions*?
14. Describe an *ideal paramagnet*!
15. What is the equation of state of the ideal paramagnet?
16. Which connection between M, T and H is mediated by the Curie law?

17. How does the equation of state of the Weiss ferromagnet read?
18. Formulate the Curie-Weiss law!

To Sect. 1.5

1. List possible forms of energy changes in a thermodynamic system.
2. Present examples for the differential work δW.
3. When is a differential form δA a total differential?
4. What is an *integrating factor*?
5. Why is the energy form *work* not a state quantity?
6. Find arguments for the expressions $\delta W = -p\, dV$ and $\delta W = \mathbf{B}_0 \cdot d\mathbf{m}$!
7. What do we understand by *magnetization work*?

Chapter 2
Laws of Thermodynamics

2.1 First Law of Thermodynamics, Internal Energy

It is extremely difficult to establish the term **heat** within the framework of the phenomenological thermodynamics with a sufficient degree of precision. We will manage that very much smoother by the means and methods of Statistical Mechanics. In thermodynamics we are forced in a certain sense to accept an **intuitive self-concept** of this term.

The first law of thermodynamics, which we are going to formulate in this section, makes a statement about the nature of heat. Experience shows that the temperature of a system can be changed without performing work on it in the sense defined in the last chapter. An essential element of the **first law of thermodynamics** consists therefore in the statement:

'heat' is a special form of energy

The system receives and gives off, respectively, this form of energy when it changes its temperature without any work is done on it or by it.

The kinetic theory of gases interprets *heat* as the energy of motion of the gas molecules, where the distinction to the kinetic energy of macroscopic bodies lies in the *disorder*. An example may help to clarify that. When a balloon filled with gas is moving around, we interpret the energy of motion of the center of gravity as the kinetic energy of the macroscopic system. But then there is still the disordered movement of the gas molecules within the balloon, which is to be understood as heat. An essential characteristic of this type of energy is thus the disorder. Hence it is reasonably definable only for many-particle systems.

© Springer International Publishing AG 2017
W. Nolting, *Theoretical Physics 5*, DOI 10.1007/978-3-319-47910-1_2

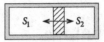

Fig. 2.1 Isolated system consisting of two partial systems, between which heat and energy, respectively, can be exchanged

When we therefore, backed by empirical facts, postulate that there does exist an independent energy form *'heat'*, and when we further assume that it is, like any other type of energy, an extensive variable then we can use:

$$dE_H = T\,dS\,.$$

T is an intensive and S an extensive quantity. E_H is the *energy of heat*. The *extensive variable S* we will later call *entropy*. It will define, in the final analysis, the energy form *heat*.

We consider an isolated system that consists of two partial systems between which an exchange of S and E_H is possible (Fig. 2.1). The total entropy $S = S_1 + S_2$ is spread over both the systems such that the energy of the total system becomes minimal (empirical fact!). At equilibrium $E_H = E_H^{(1)} + E_H^{(2)}$ is minimal at $S = S_1 + S_2 = $ const. That means:

$$0 \stackrel{!}{=} \frac{dE_H}{dS_1} = \frac{dE_H^{(1)}}{dS_1} + \frac{dE_H^{(2)}}{dS_1} = \frac{dE_H^{(1)}}{dS_1} + \frac{dE_H^{(2)}}{dS_2}\frac{dS_2}{dS_1}$$

$$= \frac{dE_H^{(1)}}{dS_1} - \frac{dE_H^{(2)}}{dS_2} = T_1 - T_2\,.$$

At equilibrium the two systems have the same T. The prefactor in the above ansatz for dE_H hence has exactly the property which we ascribe, according to the zeroth law of thermodynamics, to the term *'temperature'*.

The first law of thermodynamics, which postulates heat to be a special form of energy, has to be still brought into a mathematical form. For this purpose we introduce a new state variable,

U: **internal energy,**

which represents the total energy content of the system. It must be a unique function of the independent state variables, e.g. T and V. If it were namely possible for the system to come along two different paths from state A to the state B (Fig. 2.2), where the energy changes $\Delta U_{AB}^{(1)}$, $\Delta U_{AB}^{(2)}$ are different, e.g. $\Delta U_{AB}^{(1)} < \Delta U_{AB}^{(2)}$, then one would go along the path (1) from A to B with the expenditure of $\Delta U_{AB}^{(1)}$ and would regain more energy on the way back (2) than the one needed on the forward run. One would have created therewith energy from next to nothing (**perpetuum**

Fig. 2.2 Justification for the internal energy to be a state quantity

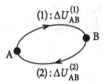

mobile of the first kind).—For a thermodynamic cycle it must in fact hold:

$$\oint dU = 0 . \tag{2.1}$$

dU is thus a total differential!

After these preparations we are now able to formulate mathematically the first law of thermodynamics which is nothing else but the energy law:

Theorem 2.1.1 (First Law of Thermodynamics)

1) Isolated systems

$$dU = 0 . \tag{2.2}$$

2) Closed systems

$$dU = \delta Q + \delta W . \tag{2.3}$$

*We use for the heat the usual letter Q. δQ is, like δW, **not** a total differential.*

$$\delta Q \; : \; energy \; due \; to \; heat\text{-}exchange \; ,$$

$$\delta W \; : \; energy \; due \; to \; work\text{-}exchange \; .$$

3) Open systems

$$dU = \delta Q + \delta W + \delta E_C . \tag{2.4}$$

Thereby it holds:

$$\delta E_C = \sum_{i=1}^{\alpha} \mu_i \, dN_i , \tag{2.5}$$

δE_C *energy due to particle-exchange,*

$N_{i, i=1,...,\alpha}$ *number of particles of the kind i,*

μ_i **chemical potential.** *That is the energy, which is needed to add an additional particle of the kind i to the system when $\delta W = \delta Q = 0$.*

We can understand the state quantity U as independent variable or, contrawise, as state function of other independent variables, e.g.:

$$U = U(T, V, N) \qquad \textbf{caloric} \text{ state equation}$$

or

$$U = U(T, p, N) , \quad U = U(V, p, N) , \quad \dots$$

One calls the relation

$$p = p(T, V, N) ,$$

in contrast to $U = U(T, V, N)$, the **thermal** state equation.

It is not the job of thermodynamics to derive the concrete form of the internal energy for special physical systems. We therefore accept here, in each case without proof the corresponding expressions. Let us mention three examples:

1) Ideal gas

$$U = U(T) , \quad \text{independent of } V . \tag{2.6}$$

This is the result of the experiment of Gay-Lussac.

$$U = \tfrac{3}{2} N k_{\mathrm{B}} T : \text{monoatomic gas molecules} ,$$
$$U = \tfrac{5}{2} N k_{\mathrm{B}} T : \text{diatomic gas molecules} ,$$
$$U = 3 N k_{\mathrm{B}} T : \text{spatial molecules} .$$

2) Solid

The following strongly simplified expression is sufficient for many purposes at high temperatures:

$$U = U_V (T) + U_{\mathrm{el}}(V) ,$$
$$U_V (T) = 3 N k_{\mathrm{B}} T , \tag{2.7}$$
$$U_{\mathrm{el}}(V) = \frac{1}{2\kappa} \frac{(V - V_0)^2}{V_0} .$$

κ is the *compressibility*.

3) Black body radiator (photon gas)

$$U = V \varepsilon(T) ; \quad p = \frac{1}{3} \varepsilon(T) . \tag{2.8}$$

The energy density $\varepsilon(T)$ is a function of temperature, only.

2.2 Heat Capacities

Heat capacities indicate the amount of change of temperature dT which is a reaction of the system to a differential heat supply δQ. Since in addition to the temperature T, there are still further state variables, we have to first clarify how these variables behave during the change of the state.

Definition 2.2.1 (Heat Capacity)

$$C_x = \left(\frac{\delta Q}{dT}\right)_x .$$
(2.9)

x stands for one or more state variables which are kept constant during the heat supply δQ.

Definition 2.2.2 (Specific Heat)

$$\bar{c}_x = \left(\frac{\delta Q}{M\, dT}\right)_x ; \quad M : \text{ mass of the system }.$$
(2.10)

Definition 2.2.3 (Molecular Heat (Molar Heat Capacity))

$$C_x^{\mathrm{mol}} = \left(\frac{\delta Q}{n\, dT}\right)_x ; \quad n : \text{ number of moles }.$$
(2.11)

We presume a closed system ($N_i = $ const) whose internal energy U in general depends on the temperature T and the generalized coordinates q_i:

$$U = U(T, q_1, \ldots, q_m) .$$

We solve the first law of thermodynamics in the form (2.3) for δQ using (1.32):

$$\delta Q = dU - \sum_{i=1}^{m} F_i\, dq_i$$

$$= \left(\frac{\partial U}{\partial T}\right)_{\mathbf{q}} dT + \sum_{i=1}^{m} \left[\left(\frac{\partial U}{\partial q_i}\right)_{T, q_{j,j \neq i}} - F_i\right] dq_i .$$
(2.12)

From this equation we read off the following special cases:

1. $\{q_i\} = \mathbf{q} = $ const
   All dq_i are then equal to zero so that:

$$C_{\mathbf{q}} = \left(\frac{\delta Q}{dT}\right)_{\mathbf{q}} = \left(\frac{\partial U}{\partial T}\right)_{\mathbf{q}} .$$
(2.13)

2. $\{F_i\} = \mathbf{F} = \text{const}$

   At first the equations of state

$$F_j = F_j(q_1, \ldots, q_m, T) \; ; \quad j = 1, \ldots, m$$

must be solved for q_i:

$$q_i = q_i(F_1, \ldots, F_m, T)$$

$$\Longrightarrow dq_i = \sum_{j=1}^{m} \left(\frac{\partial q_i}{\partial F_j} \right)_{T, F_{k, k \neq j}} dF_j + \left(\frac{\partial q_i}{\partial T} \right)_{\mathbf{F}} dT \, .$$

This leads to the heat capacity:

$$C_{\mathbf{F}} = \left(\frac{\delta Q}{dT} \right)_{\mathbf{F}} = \left(\frac{\partial U}{\partial T} \right)_{\mathbf{q}} + \sum_{i=1}^{m} \left[\left(\frac{\partial U}{\partial q_i} \right)_{T, q_{j, j \neq i}} - F_i \right] \left(\frac{\partial q_i}{\partial T} \right)_{\mathbf{F}} \, . \qquad (2.14)$$

We discuss some important **examples**:

1) Gas

$$q = V \; ; \quad F = -p \, .$$

It then follows according to (2.13):

$$C_V = \left(\frac{\delta Q}{dT} \right)_V = \left(\frac{\partial U}{\partial T} \right)_V \, . \qquad (2.15)$$

Equation (2.14), on the other hand, reads:

$$C_p = \left(\frac{\delta Q}{dT} \right)_p = \left(\frac{\partial U}{\partial T} \right)_V + \left[\left(\frac{\partial U}{\partial V} \right)_T + p \right] \left(\frac{\partial V}{\partial T} \right)_p \, . \qquad (2.16)$$

This yields:

$$C_p - C_V = \left[\left(\frac{\partial U}{\partial V} \right)_T + p \right] \left(\frac{\partial V}{\partial T} \right)_p \, . \qquad (2.17)$$

Special case: ideal gas

$$\left(\frac{\partial U}{\partial V} \right)_T \overset{(2.6)}{=} 0 \; ; \quad \left(\frac{\partial V}{\partial T} \right)_p = \frac{nR}{p}$$

$$\Longrightarrow C_p - C_V = nR = N k_B \, . \qquad (2.18)$$

Thus we must have $C_p > C_V$.

2) Magnet

$$q = m ; \quad F = B_0 = \mu_0 H .$$

Equation (2.13) now reads:

$$C_m = \left(\frac{\delta Q}{dT}\right)_m = \left(\frac{\partial U}{\partial T}\right)_m . \tag{2.19}$$

We derive from (2.14):

$$C_H - C_m = \left[\left(\frac{\partial U}{\partial m}\right)_T - \mu_0 H\right]\left(\frac{\partial m}{\partial T}\right)_H . \tag{2.20}$$

2.3 Adiabatics, Isotherms

Let us discuss special kinds of state changes on the basis of the first law of thermodynamics. These are characterized by the fact that, when performed, certain independent or dependent state quantities are to be kept constant.

Adiabatic state changes are defined by

$$\delta Q = 0 .$$

We will mark them by the index 'ad'. The state function, which is constant during such processes, is the entropy S, which we will get to know later in this course.

Starting point is the first law of thermodynamics in the form of (2.12):

$$\left(\frac{\partial U}{\partial T}\right)_\mathbf{q} (dT)_{ad} = \sum_{i=1}^{m} \left[F_i - \left(\frac{\partial U}{\partial q_i}\right)_{T, q_{j,j \neq i}}\right] (dq_i)_{ad} . \tag{2.21}$$

This we investigate in more detail for some standard examples:

1) Gas

$$q = V , \quad F = -p$$

$$\implies \left(\frac{\partial U}{\partial T}\right)_V (dT)_{ad} = -\left[p + \left(\frac{\partial U}{\partial V}\right)_T\right] (dV)_{ad} .$$

This yields:

$$\left(\frac{dT}{dV}\right)_{ad} = -\frac{p + \left(\frac{\partial U}{\partial V}\right)_T}{C_V} . \tag{2.22}$$

Special case: ideal gas

$$\left(\frac{\partial U}{\partial V}\right)_T = 0 \implies \left(\frac{dT}{dV}\right)_{\text{ad}} = -\frac{p}{C_V} = -\frac{nRT}{C_V V} \ .$$

With (2.18) we get further on:

$$\left(\frac{dT}{T}\right)_{\text{ad}} = -\frac{C_p - C_V}{C_V}\left(\frac{dV}{V}\right)_{\text{ad}} \ .$$

One defines

$$\gamma = \frac{C_p}{C_V} \tag{2.23}$$

and obtains therewith:

$$(d \ln T)_{\text{ad}} = -(\gamma - 1)\,(d \ln V)_{\text{ad}} \implies \left(d \ln T\, V^{\gamma - 1}\right)_{\text{ad}} = 0 \ .$$

This means eventually:

$$T\,V^{\gamma - 1} = \text{const}_1 \ . \tag{2.24}$$

Inserting the equation of state of the ideal gas leads to two further **adiabatic equations**:

$$p\,V^{\gamma} = \text{const}_2 \ ; \quad T^{\gamma}\,p^{1-\gamma} = \text{const}_3 \ . \tag{2.25}$$

2) Black body radiator

By a *black body radiator* one understands the electromagnetic radiation field, which adjusts itself in thermal equilibrium in a hollow space of the volume V, which is enclosed by a heat bath with the temperature T (Fig. 2.3). The electromagnetic radiation is thereby emitted by the hollow walls (*heat radiation*). One can show that its energy density $\varepsilon(T)$ is solely a function of temperature so that for the internal energy U we can use (2.8):

$$U(T, V) = V\,\varepsilon(T) \ .$$

Fig. 2.3 Simple scheme of a
black body radiator

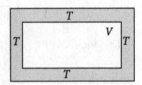

The connection between radiation pressure p and energy density $\varepsilon(T)$ in the isotropic radiation field,

$$p = \frac{1}{3} \varepsilon(T) \,,$$

can be verified in the framework of classical electrodynamics (see Exercise 4.3.2, Vol. 3).

From atomic physics we know that radiation of a certain frequency ν appears only in discrete energy quanta

$$\varepsilon_\nu = h \nu \,.$$

This leads to the concept of the **photon**, which one can visualize in an illustrative manner as **quasiparticle** with the energy $h\nu$, the momentum $(h\nu)/c$, the velocity c, and the mass $m = 0$. The radiation field in V can therefore also be interpreted as a **gas of photons**, which obeys the laws of the kinetic theory of gases. The above relation for the radiation pressure is in this sense easily derived as the momentum transfer of the photons onto the walls of the hollow space. (Show it!)

The heat capacity of the photon gas reads:

$$C_V = \left(\frac{\partial U}{\partial T}\right)_V = V \frac{d\varepsilon}{dT} \,. \tag{2.26}$$

For C_p we would have to calculate according to (2.17), among others, the term $\left(\frac{\partial V}{\partial T}\right)_p$. Since $p = $ const automatically entails $T = $ const, this expression is not defined. The photon gas has no C_p.

The adiabatic equation (2.22) yields for the black body radiator:

$$\left(\frac{dT}{dV}\right)_{ad} = -\frac{\frac{1}{3}\varepsilon(T) + \varepsilon(T)}{V \frac{d\varepsilon}{dT}}$$

$$\Longrightarrow \quad -\frac{d\varepsilon}{\varepsilon} = \frac{4}{3}\frac{dV}{V} \quad \Longleftrightarrow \quad d\ln\left(V^{4/3}\,\varepsilon\right) = 0 \,.$$

Finally that results in:

$$\varepsilon\, V^{4/3} = \text{const}_4 \,; \quad p\, V^{4/3} = \text{const}_5 \,. \tag{2.27}$$

Isothermal state changes are defined by

$$dT = 0 \,.$$

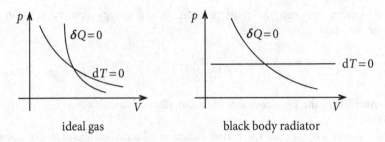

Fig. 2.4 Isotherms and adiabatics of the ideal gas (*left*) and the black radiator (*right*)

The first law of thermodynamics in the form (2.12) in this case reads:

$$(\delta Q)_T = \sum_{i=1}^{m} \left[\left(\frac{\partial U}{\partial q_i} \right)_{T, q_{j,j \neq i}} - F_i \right] (dq_i)_T \; . \tag{2.28}$$

That means for a **gas** with $q = V$ and $F = -p$:

$$\left(\frac{\delta Q}{dV} \right)_T = \left(\frac{\partial U}{\partial V} \right)_T + p \; . \tag{2.29}$$

1) Ideal gas

$$\left(\frac{\partial U}{\partial V} \right)_T = 0 \implies (\delta Q)_T = (p \, dV)_T \; . \tag{2.30}$$

2) Photon gas

$$\left(\frac{\delta Q}{dV} \right)_T = \frac{4}{3} \varepsilon(T) = \text{const} \; . \tag{2.31}$$

Adiabatic and isothermal state changes in the pV-diagram exhibit, qualitatively, a behavior such as sketched in Fig. 2.4.

Because $\gamma > 1$ the adiabatic curve, in the case of the ideal gas, drops steeper than the isotherm.

2.4 Second Law of Thermodynamics

It is obvious that the first law of thermodynamics does not suffice for a complete description of thermodynamic systems. One can easily think of physical processes, which are absolutely allowed from the point of view of conservation of energy, but have never been observed in nature:

1. Why has it never been observed that a stone, lying on the ground, jumps onto the rooftop taking the necessary energy for the jump simply by cooling-down?
2. Why can an ocean liner not move without any special power supply, simply by changing heat from the huge water reservoir into mechanical work, where part of it would even be given back to the ocean in form of frictional heat?

Experience tells us that a series of energy conversions, for which heat comes into play, are not reversible. We know that work can completely be converted, e.g. by friction, into heat. Think, for instance, of a macroscopic body gliding on a rough base and being set in motion by an initial momentum. After a finite time the body comes to rest. Mechanical work has been changed into heat by friction. The reversal, namely that the resting body starts moving again by cooling-down, is according to the first law definitely allowed but never happens. If this inverse process existed then one would have had a

perpetuum mobile of the second kind:

That is a **periodically (cyclically)** working thermodynamic machine, which does nothing else but executing work per cycle, where a heat amount ΔQ is taken only from one single heat reservoir.

Theorem 2.4.1 (Second Law of Thermodynamics) *A perpetuum mobile of the second kind does not exist!*

In thermodynamics this theorem is accepted without strict proof as **never being contradicted empirical fact**.

The above version of the second law of thermodynamics is called **Kelvin's statement**. It expresses that there cannot be a state change, whose **only** effect consists of having drawn a certain amount of heat from a heat reservoir and converted completely into work.

There exists an equivalent formulation:

Clausius' Statement *There does not exist a **periodically** working machine which does **nothing else but** take heat from a colder heat bath and transfer it to a hotter heat bath.*

The key-words of this statement are strictly to be taken into consideration:

$$periodic \iff \text{cyclic process,}$$
$$nothing\ else \iff \text{apart from that nothing happens,}$$
$$\text{in the surroundings, either.}$$

In this context we introduce a new term.

Definition 2.4.1 (Heat Engine) That is a thermodynamic system which performs a cyclic process between two heat baths $HB(T_1)$ and $HB(T_2)$ with $T_1 > T_2$, where exactly the following happens:

1. $\Delta Q_1 > 0$ by contact to $HB(T_1)$,
2. $\Delta W < 0$,
3. $\Delta Q_2 < 0$ by contact to $HB(T_2)$.

Such machines do **not** violate the second law of thermodynamics because they are in contact with **two** heat baths, where the heat that is drawn from the first bath is not completely converted into work. We must have $|\Delta Q_2| < |\Delta Q_1|$, since the first law must also be fulfilled. One ascribes to such a machine a degree of efficiency:

Definition 2.4.2 ((Energy Conversion) Efficiency)

$$\eta = \frac{\text{work done by the system}}{\text{amount of heat absorbed}} = \frac{-\Delta W}{\Delta Q_1}. \tag{2.32}$$

Let us prove the equivalence of the two formulations of the second law of thermodynamics.

Proposition 1 *If the Clausius' statement is wrong, then the Kelvin's statement is also wrong.*

a) With a periodically working machine we take $\Delta Q_1 > 0$ from the heat bath $HB(T_2)$ and put it into the bath $HB(T_1)$ where $T_1 > T_2$. This is possible because the Clausius' statement is assumed to be wrong.
b) We run a heat engine such that ΔQ_1 is taken from $HB(T_1)$ and then $\Delta Q_2 < 0$ ($|\Delta Q_2| < \Delta Q_1$) is given back to $HB(T_2)$ with the work expenditure $\Delta W < 0$.

On the whole $\Delta Q = \Delta Q_1 + \Delta Q_2 > 0$ from $HB(T_2)$ has been completely converted into work. Otherwise, nothing has happened since a) as well as b) are cyclic processes. Therewith Kelvin's statement is also wrong!

Proposition 2 *If the Kelvin's statement is wrong, then the Clausius' statement is also wrong.*

a) We take $\Delta Q > 0$ from the heat bath $HB(T_2)$ and convert it completely with a periodically working engine into work. That is possible since Kelvin's statement is assumed to be wrong.
b) We convert the work from a) completely into heat. That is always possible, only the reverse process is not. We put the so obtained heat into $HB(T_1)$, where $T_1 > T_2$.

On the whole $\Delta Q > 0$ is exclusively transferred from $HB(T_2)$ into $HB(T_1)$ in spite of $T_1 > T_2$. The Clausius' statement is therewith also wrong!

By combining the two propositions we recognize the equivalence of the statements of Clausius and Kelvin.

2.5 Carnot Cycle

In a *thermodynamic cycle*, the system runs through various (heat, work, particle)-exchange contacts, finally coming back to its initial state. Please note that only the thermodynamic system returns to its initial state, the **surroundings**, however, might certainly change. For instance, energy in the form of work and heat could have been interchanged between different *reservoirs*. According to the first law of thermodynamics it holds of course

$$0 = \oint dU = \oint \delta Q + \oint \delta W ,$$

but the two terms on the right-hand side can be unequal zero (opposite and equal)!

We are now going to discuss in detail a very special thermodynamic cycle, i.e., a very special *heat engine*.

Carnot Cycle   That is a **reversible** thermodynamic cycle built up by two adiabatics and two isotherms between two heat baths $HB(T_1)$ and $HB(T_2)$ with $T_1 > T_2$. It consists of the following segments (Fig. 2.5):

- $a \to b$
  Adiabatic compression with

$$\Delta T = T_1 - T_2 > 0 .$$

- $b \to c$
  Isothermal expansion with a heat absorption of $\Delta Q_1 > 0$ from $HB(T_1)$.
- $c \to d$
  Adiabatic expansion with $\Delta T = T_2 - T_1 < 0$.
- $d \to a$
  Isothermal compression with a heat delivery of $\Delta Q_2 < 0$ into $HB(T_2)$.

The work performed in one cycle corresponds just to the area enclosed by the path $a \to b \to c \to d \to a$.

We symbolize the Carnot cycle by the diagram in Fig. 2.6.

Fig. 2.5 Adiabatics and isotherms of the Carnot cycle in the pV-diagram

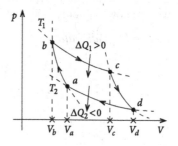

Fig. 2.6 Symbolic
representation of the Carnot
cycle as heat engine

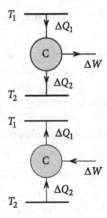

Fig. 2.7 Symbolic
representation of the Carnot
cycle as heat pump

The first law of thermodynamics requires at first:

$$0 = \oint dU = \Delta Q_1 + \Delta Q_2 + \Delta W \ .$$

This yields the efficiency of the heat engine:

$$\eta = \frac{-\Delta W}{\Delta Q_1} = \frac{\Delta Q_1 + \Delta Q_2}{\Delta Q_1} = 1 + \frac{\Delta Q_2}{\Delta Q_1} \ . \tag{2.33}$$

Because of $\Delta Q_2/\Delta Q_1 < 0$ we always have $\eta < 1$.

Since the Carnot process has to be reversible, the sense of rotation can be reversed
(Fig. 2.7):

$$\Delta Q_2 > 0 \ ; \quad \Delta Q_1 < 0 \ ; \quad \Delta W > 0$$
$$|\Delta Q_1| > \Delta Q_2 \ .$$

The machine then works as

heat pump

Let the **'working substance'** of the Carnot machine be the **ideal gas**. Therewith we
want to now explicitly calculate the efficiency of the machine.

- $a \rightarrow b$ *Adiabatic*

$$\Delta Q = 0 \iff \Delta W = \Delta U$$
$$\implies \Delta W_{ab} = C_V \ (T_1 - T_2) = -\Delta W_{cd} \ .$$

- $b \to c$ *Isotherm*

$$\Delta W_{bc} = -\int_b^c p(V)\, dV = -nRT_1 \int_{V_b}^{V_c} \frac{dV}{V}$$

$$= -nRT_1 \ln \frac{V_c}{V_b} \,. \tag{2.34}$$

- $c \to d$ *Adiabatic*
  see $(a \to b)$
- $d \to a$ *Isotherm*

$$\Delta W_{da} = -nRT_2 \ln \frac{V_a}{V_d} \,. \tag{2.35}$$

On the adiabatics it holds according to (2.24):

$$T_2 V_a^{\gamma-1} = T_1 V_b^{\gamma-1},$$

$$T_2 V_d^{\gamma-1} = T_1 V_c^{\gamma-1} \implies \frac{V_a}{V_d} = \frac{V_b}{V_c} \,.$$

Therewith one gets for the total performed work:

$$\Delta W = \Delta W_{ab} + \Delta W_{bc} + \Delta W_{cd} + \Delta W_{da}$$

$$= \Delta W_{bc} + \Delta W_{da}$$

$$\implies \Delta W = -nR(T_1 - T_2) \ln \frac{V_d}{V_a} < 0 \,. \tag{2.36}$$

On the isotherm $b \to c$ it is $\Delta U = 0$ and therewith

$$\Delta Q_1 = -\Delta W_{bc} = nRT_1 \ln \frac{V_c}{V_b} = nRT_1 \ln \frac{V_d}{V_a} > 0 \,.$$

According to the definition (2.32) we then have as the **efficiency** η_C of the **Carnot cycle**:

$$\eta_C = 1 - \frac{T_2}{T_1} \,. \tag{2.37}$$

As direct consequences of the second law of thermodynamics we now prove the following two **propositions**:

1. The Carnot process has the **highest** efficiency of all machines working periodically between two heat baths.
2. η_C is reached by all **reversibly** working machines.

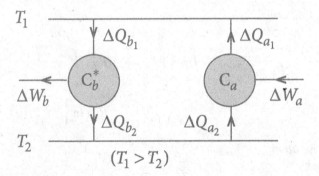

Fig. 2.8 Schematic arrangement of a Carnot machine and an arbitrary heat engine for the investigation of the efficiency of the Carnot cycle. C_a: Carnot-machine running as heat pump. C_b^*: heat engine, not necessarily reversible

Proof Consider the situation plotted in Fig. 2.8 concerning a Carnot machine C_a and an arbitrary heat engine C_b^*, both between the same heat baths $HB(T_1)$ and $HB(T_2)$. The machines shall be dimensioned such that $\Delta Q_{b_2} = -\Delta Q_{a_2} < 0$, i.e., the heat bath $HB(T_2)$ remains uninfluenced (Fig. 2.8). However, $HB(T_1)$ exchanges with the total system $C_a \cup C_b^*$ the heat

$$\Delta Q = \Delta Q_{b_1} + \Delta Q_{a_1}$$

According to the second law of thermodynamics it has to be

$$\Delta Q \le 0 \,,$$

since otherwise the system $C_a \cup C_b^*$ would do nothing else but take heat from the bath $HB(T_1)$ and change it completely into work.

$$\eta_{C_a} = \eta_C = 1 + \frac{-\Delta Q_{a_2}}{-\Delta Q_{a_1}} \iff \Delta Q_{a_1} = \Delta Q_{a_2} \frac{1}{\eta_C - 1} \,,$$

$$\eta_{C_b^*} = 1 + \frac{\Delta Q_{b_2}}{\Delta Q_{b_1}} = 1 - \frac{\Delta Q_{a_2}}{\Delta Q_{b_1}} \iff \Delta Q_{b_1} = -\Delta Q_{a_2} \frac{1}{\eta_{C_b^*} - 1} \,.$$

After insertion it follows:

$$0 \ge \Delta Q_{b_1} + \Delta Q_{a_1} = \Delta Q_{a_2} \left(\frac{1}{\eta_C - 1} - \frac{1}{\eta_{C_b^*} - 1} \right) \,.$$

Since ΔQ_{a_2} is positive, Proposition 1. is proven:

$$\eta_{C_b^*} \le \eta_C \,. \tag{2.38}$$

If C_b^* is a reversible machine then the sense of rotation in the arrangement sketched in Fig. 2.8 can be also reversed. All the above expressions retain their validity, except for the statement $\Delta Q_{a_2} > 0$, which now must be read $\Delta Q_{a_2} < 0$. For machines which act reversibly between the two heat baths it then holds in addition to (2.38) also $\eta_{C_b^*} \geq \eta_C$. Hence only the equality sign is valid. Therewith Proposition 2. also is proven.

The efficiency η_C of **reversible** thermodynamic cycles thus turns out to be universal!

2.6 Absolute, Thermodynamic Temperature Scale

We have seen that the universal degree of efficiency η_C of the Carnot cycle depends only on the temperatures of the involved heat baths $HB(T_1)$ and $HB(T_2)$ if we use the ideal gas as 'working substance'. Thereby we remember that we have introduced the temperature T itself in (1.3) and (1.5), respectively, through the equation of state of the ideal gas. It is of course somewhat unfortunate that, as a basis for the definition of such an important concept as *temperature*, we to take recourse to a system which hardly exists in a strict sense. On top of that, we have to drive with this system a machine, by which we want to derive quite a wealth of far reaching conclusions.

It turns out, however, that we can also utilize **inversely** the universal efficiency η_C of the Carnot cycle in order to **define** the temperatures ϑ_1, ϑ_2 of the involved heat baths. That succeeds because the proof of the universality of the efficiency of reversible thermodynamic cycles, as we performed it in the last section, did not at all need the presumption of an *ideal gas*, but resulted very generally from the second law of thermodynamics.—Since, on the other hand, η_C as the ratio of two amounts of energy is directly and readily measurable, we are now going to introduce via η_C a

universal, substance independent, thermodynamic temperature scale.

Let ϑ be an arbitrarily chosen temperature scale which only guarantees that we have:

$$\text{'warmer'} \iff \text{'larger'} \vartheta \ .$$

We consider three heat baths $HB(\vartheta_1)$, $HB(\vartheta_2)$ and $HB(\vartheta_3)$ with $\vartheta_1 > \vartheta_2 > \vartheta_3$ (Fig. 2.9). Let C_a, C_b be any two heat engines, reversibly working, respectively, between $HB(\vartheta_1)$ and $HB(\vartheta_2)$ and between $HB(\vartheta_2)$ and $HB(\vartheta_3)$. The machine C_b may be designed such that

$$\Delta Q_{b_2} = -\Delta Q_{a_2} \ .$$

Fig. 2.9 Schematic
combination of Carnot cycles
for fixing an absolute
substance independent
temperature scale

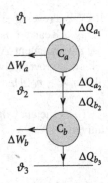

To $HB(\vartheta_2)$, all in all, it does not happen anything. The efficiencies of the two
machines

$$\eta_{C_a} = 1 + \frac{\Delta Q_{a_2}}{\Delta Q_{a_1}} \, ,$$

$$\eta_{C_b} = 1 + \frac{\Delta Q_{b_3}}{\Delta Q_{b_2}}$$

are universal. That means any other reversible machine would deliver the same
efficiency. Furthermore, the efficiencies are independent of the working substance.
If, however, the type of the machine does not play any role then the efficiencies can
depend only on the *temperatures* ϑ_i of the heat baths. For the above system there are
no other discriminating characteristics. The following ansatz-functions are therefore
reasonable:

$$\eta_{C_a} = 1 - f(\vartheta_1, \vartheta_2) \, ,$$
$$\eta_{C_b} = 1 - f(\vartheta_2, \vartheta_3) \, .$$

Since the machines are designed in such a manner that in the end $HB(\vartheta_2)$ remains
inactive, we can consider the total system also as a single machine reversibly
running between $HB(\vartheta_1)$ and $HB(\vartheta_3)$:

$$\eta_{C_{ab}} = 1 - f(\vartheta_1, \vartheta_3) \, .$$

For the work outputs we have therewith:

$$-\Delta W_a = \Delta Q_{a_1} (1 - f(\vartheta_1, \vartheta_2)) \, ,$$
$$-\Delta W_b = \Delta Q_{b_2} (1 - f(\vartheta_2, \vartheta_3)) \, ,$$
$$-\Delta W_{ab} = \Delta Q_{a_1} (1 - f(\vartheta_1, \vartheta_3)) \, .$$

Furthermore we have:

$$\Delta Q_{b_2} = -\Delta Q_{a_2} = -\Delta Q_{a_1} (\eta_{C_a} - 1) = \Delta Q_{a_1} f (\vartheta_1, \vartheta_2) .$$

Further, exploiting

$$\Delta W_{ab} = \Delta W_a + \Delta W_b$$

we get:

$$\left(1 - f (\vartheta_1, \vartheta_3)\right) = \left(1 - f (\vartheta_1, \vartheta_2)\right) + f (\vartheta_1, \vartheta_2) \left(1 - f (\vartheta_2, \vartheta_3)\right) .$$

This yields the following conditional equation:

$$f (\vartheta_1, \vartheta_3) = f (\vartheta_1, \vartheta_2) f (\vartheta_2, \vartheta_3) . \tag{2.39}$$

Because of

$$\ln f (\vartheta_1, \vartheta_3) = \ln f (\vartheta_1, \vartheta_2) + \ln f (\vartheta_2, \vartheta_3)$$

it must then also be valid that

$$\frac{\partial}{\partial \vartheta_1} \ln f (\vartheta_1, \vartheta_3) = \frac{\partial}{\partial \vartheta_1} \ln f (\vartheta_1, \vartheta_2) .$$

But this can be correct only if f can be written as follows:

$$f (\vartheta_1, \vartheta_2) = \alpha (\vartheta_1) \beta (\vartheta_2) .$$

This is inserted into (2.39):

$$\alpha (\vartheta_1) \beta (\vartheta_3) = \alpha (\vartheta_1) \beta (\vartheta_2) \alpha (\vartheta_2) \beta (\vartheta_3)$$
$$\Longleftrightarrow 1 = \alpha (\vartheta_2) \beta(\vartheta_2) \Longleftrightarrow \alpha (\vartheta) = \beta^{-1} (\vartheta) .$$

That means for f

$$f (\vartheta_1, \vartheta_2) = \frac{\beta (\vartheta_2)}{\beta (\vartheta_1)}$$

and therewith for the efficiency:

$$\eta_{C_a} = 1 - \frac{\beta (\vartheta_2)}{\beta (\vartheta_1)} . \tag{2.40}$$

$\beta(\vartheta)$ is here at first still a completely arbitrary function. This expression is formally identical to the η_C, which we had found in (2.37) when using the ideal gas as working substance. $\beta(\vartheta)$ is uniquely determined if we ascribe to one special heat bath the value

$$T^* = \beta\left(\vartheta^*\right) .$$

Then each reversibly working machine yields uniquely the temperature quotient T/T^*. One agrees upon:

$$T^* = 273.16\,\text{K} : \quad \text{triple point of water.} \tag{2.41}$$

$T = \beta(\vartheta)$ defines therewith an absolute, substance independent temperature

$$T = T^*\left(1 - \eta_C\left(T^*, T\right)\right) , \tag{2.42}$$

which turns out to be identical to the so far applied ideal gas temperature.

2.7   Entropy as State Quantity

The methods of deduction used so far, which are all based upon the second law of thermodynamics, now allow us to introduce the for thermodynamics certainly the most important quantity, namely the **entropy**.

We had found for the efficiency of the Carnot cycle:

$$\eta_C = 1 - \frac{T_2}{T_1} = 1 + \frac{\Delta Q_2}{\Delta Q_1} .$$

That means:

$$\frac{\Delta Q_1}{T_1} + \frac{\Delta Q_2}{T_2} = 0 . \tag{2.43}$$

Let us now further generalize this result.

A thermodynamic system moves quasi-statically along a (not necessarily reversible) cycle K. For the description of the temperature change we decompose the cycle into n steps (Fig. 2.10). During each step the temperature of the system is kept constant by its contact with a heat bath

$$HB\left(T_i\right) ; \quad i = 1, 2, \ldots, n .$$

Fig. 2.10 Thermodynamic cycle in contact with n heat baths of different temperatures

Fig. 2.11 Thermodynamic cycle in contact with n heat baths of different temperatures T_i, where to each heat bath a Carnot machine is coupled, which works between T_i and the fixed temperature T_0. (Proof of the Clausius inequality)

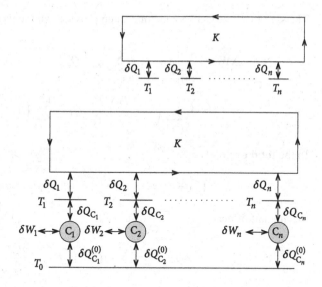

Thereby a heat exchange δQ_i takes place which can be positive as well as negative. According to the first law of thermodynamics we have then for the total work output on K:

$$\Delta W_K = -\sum_{i=1}^{n} \delta Q_i \, .$$

We now couple to each $HB(T_i)$ a Carnot machine C_i, which works between this $HB(T_i)$ and a heat bath $HB(T_0)$, where

$$T_0 > T_i \qquad \forall i \, .$$

Each C_i is capable of working either as a heat engine or as a heat pump (Fig. 2.11). We arrange the C_i in such a way that they absorb just the amount of heat from $HB(T_i)$, which was given from the considered thermodynamic system to $HB(T_i)$ (or vice versa):

$$\delta Q_{C_i} = -\delta Q_i \qquad \forall i \, .$$

For each Carnot machine we have:

$$\delta Q_{C_i}^{(0)} = -\frac{T_0}{T_i} \delta Q_{C_i} = \frac{T_0}{T_i} \delta Q_i \, .$$

The system of the Carnot machines then provides on the whole the work:

$$\Delta W_C = \sum_{i=1}^{n} \delta W_i = -\sum_{i=1}^{n} \eta_{C_i}\, \delta Q_{C_i}^{(0)}$$

$$= -\sum_{i=1}^{n} \left(1 - \frac{T_i}{T_0}\right) \frac{T_0}{T_i} \delta Q_i = \sum_{i=1}^{n} \left(1 - \frac{T_0}{T_i}\right) \delta Q_i \,.$$

In the **total cycle**

$$K + \{C_1 + C_2 + \ldots + C_n\} \quad (thermodynamic\ cycle)$$

the amount of heat

$$\Delta Q^{(0)} = \sum_{i=1}^{n} \delta Q_{C_i}^{(0)} = T_0 \sum_{i=1}^{n} \frac{\delta Q_i}{T_i} \tag{2.44}$$

is exchanged with $HB(T_0)$ and thereby the work

$$\Delta W = \Delta W_K + \Delta W_C = -T_0 \sum_{i=1}^{n} \frac{\delta Q_i}{T_i} \tag{2.45}$$

is done. Except that, nothing else takes place. The first law of thermodynamics is obviously fulfilled:

$$\Delta W = -\Delta Q^{(0)} \,.$$

The second law of thermodynamics, however, requires that

$$\Delta W \geq 0. \tag{2.46}$$

In the reverse case, nothing else would have taken place except that the total thermodynamic system would have taken heat $\Delta Q^{(0)}$ from $HB(T_0)$ and converted it completely into work $\Delta W \leq 0$. That, however, is impossible. It follows therewith from (2.45) and (2.46) the important result

$$\sum_{i=1}^{n} \frac{\delta Q_i}{T_i} \leq 0 \,, \tag{2.47}$$

which contains only the *data* of the original cycle K. If this cycle is further reversible, then the direction sense of K can be reversed. As to the above considerations nothing will change at all. However, the quantities δQ_i in (2.47) change their sign. But since (2.47) is equally correct for both the senses only the equal sign does not lead to a contradiction:

$$\sum_{i=1}^{n} \frac{\delta Q_i}{T_i} = 0 \iff K \text{reversible} . \tag{2.48}$$

By generalization to $n \to \infty$ partial steps one finds with (2.47) and (2.48) the fundamental
Clausius inequality

$$\oint \frac{\delta Q}{T} \leq 0 . \tag{2.49}$$

For reversible processes it even holds:

$$\oint \frac{\delta Q_{rev}}{T} = 0 . \tag{2.50}$$

This relation defines a state quantity. Let

$$A_0 \text{ be a fixed point of the state space,}$$

then the integral

$$\int_{A_0}^{A} \frac{\delta Q_{rev}}{T}$$

is independent of the path on which we go in the state space from the state A_0 to the state A. For fixed A_0 it is a unique function of the state A. The so-called **entropy** S,

$$S(A) = \int_{A_0}^{A} \frac{\delta Q_{rev}}{T} , \tag{2.51}$$

is thus a state quantity, except for an undetermined additive constant, with the **total differential**

$$dS = \frac{\delta Q_{rev}}{T} . \tag{2.52}$$

$1/T$ is therewith the integrating factor (1.34), which makes the non-integrable differential form δQ a total differential (see Exercise 2.9.1).

One has to bear in mind the important fact, that the entropy is **always to be calculated** along a reversible path from A_0 to A. Thereby, it is besides the point

Fig. 2.12 Path of a not
necessarily reversible state
change Z, referred to a
reversible auxiliary process R

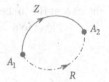

how the system **actually reaches** the state A, whether reversibly or irreversibly
(Fig. 2.12). For the determination of $S(A)$ one thus needs always a *'reversible
auxiliary process'*. For an **arbitrary** state change we have:

$$S(A_2) - S(A_1) \geq \int_{\substack{A_1 \\ (Z)}}^{A_2} \frac{\delta Q}{T} .$$

(2.53)

Proof Let R be a reversible auxiliary process (Fig. 2.12), on which it holds:

$$S(A_2) - S(A_1) = \int_{\substack{A_1 \\ (R)}}^{A_2} \frac{\delta Q}{T} .$$

Since the path R is reversible, it can be reversed and combined with Z to a thermodynamic cycle for which it must hold, according to the Clausius inequality (2.49):

$$\int_{\substack{A_1 \\ (Z)}}^{A_2} \frac{\delta Q}{T} + \int_{\substack{A_2 \\ (-R)}}^{A_1} \frac{\delta Q}{T} \leq 0 \iff -\int_{\substack{A_2 \\ (-R)}}^{A_1} \frac{\delta Q}{T} \geq \int_{\substack{A_1 \\ (Z)}}^{A_2} \frac{\delta Q}{T}$$

$$\iff S(A_2) - S(A_1) \geq \int_{\substack{A_1 \\ (Z)}}^{A_2} \frac{\delta Q}{T} \qquad \text{q. e. d.}$$

We had to presume for the derivation of the results (2.49) to (2.53) only the validity
of the second law of thermodynamics. Conversely, we therefore get from these
results a
mathematical formulation of the second law of thermodynamics

$$dS \geq \frac{\delta Q}{T} \qquad \text{(equality sign for reversible processes!)}$$

(2.54)

If one combines the first and the second law of thermodynamics one gets the
basic relation of thermodynamics

$$T\, dS \geq dU - \delta W - \delta E_C \, . \tag{2.55}$$

With this basic relation, with the definition of the entropy as a new state quantity (2.51) as well as with the introduction of the thermodynamic temperature (2.42)
the central terms of thermodynamics are now established. The following considerations represent, more or less, conclusions from this basic concept.
   Let us consider as the first **special case** an

$$\text{isolated system:} \quad dS \geq 0 \, . \tag{2.56}$$

The isolated system can not, by definition, exchange heat with the surroundings. As
long as in such a system (irreversible) processes still can take place, the entropy can
only increase. It is therefore **maximal** in the equilibrium state. This transition into
equilibrium is irreversible. Entropy-increase without any exchange characterizes
irreversible processes. We try to illustrate the physical meaning of the entropy with
a simple example:

isothermal expansion of the ideal gas.

1)   Reversible

   The gas is pushing a piston which is affixed by a spring to a wall (Fig. 2.13). The
   work that is done by the gas, when pushing the piston, is stored in the spring and
   can in principle serve to undo the shift of the piston. The expansion of the gas is
   therewith reversible.—Let the gas be in contact with a heat bath $HB(T)$ so that
   all the changes of the state changes take place isothermally:

$$U = U(T) \implies \Delta U = 0 \, .$$

With the first law of thermodynamics we then have:

$$\Delta Q = -\Delta W = \int_{V_1}^{V_2} p\, dV = n R T \ln \frac{V_2}{V_1} \, .$$

Fig. 2.13 Schematic
arrangement for a reversible
expansion of the ideal gas

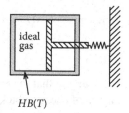

$HB(T)$

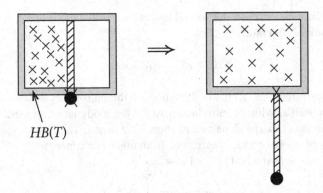

$HB(T)$

Fig. 2.14 Free expansion of the ideal gas as an example of an irreversible process

Due to this reversible change of state the entropy according to (2.54) changes:

$$(\Delta S)_{gas} = \frac{\Delta Q}{T} = n R \ln \frac{V_2}{V_1} \,.$$

The amount of heat ΔQ, needed for carrying out the work, is taken from the heat bath and can be given back to it again by compression of the gas by relieving the spring. That means that the processes in the heat bath are also reversible:

$$(\Delta S)_{HB} = \frac{-\Delta Q}{T} = -(\Delta S)_{gas} \,.$$

Hence the entropy of the total system has not changed.

2) Irreversible

The analogous irreversible process would be the **free** expansion of the ideal gas (Fig. 2.14). During the free expansion the gas does not execute work. No heat is therefore taken from the reservoir. We cannot describe the detailed variation in time of the irreversible process. Initial and end states are, however, equilibrium states. They correspond to those of the process 1). 1) is therefore a reversible auxiliary process for 2). The entropy-change is thus the same as that in 1):

$$(\Delta S)_{gas} = n R \ln \frac{V_2}{V_1} \,.$$

Because of $\Delta Q = 0$ it is, however,

$$(\Delta S)_{HB} = 0 \,.$$

According to that the entropy of the total system has increased. $T(\Delta S)_{tot}$ is just the amount of energy which in the reversible case 1) has been converted into *exploitable* work $(-\Delta W)$. That means:

Irreversibility releases exploitable energy

2.8   Simple Conclusions from the Thermodynamic Laws

We consider reversible processes in closed systems. For these the basic relation (2.55) reads as follows:

$$T\,dS = dU - \delta W . \tag{2.57}$$

A series of important conclusions are already found by the fact that dS and dU are total differentials. We think at first of T and V as independent state variables (gas!):

$$S = S(T, V) ; \quad U = U(T, V)$$

$$\implies dS = \left(\frac{\partial S}{\partial T}\right)_V dT + \left(\frac{\partial S}{\partial V}\right)_T dV = \frac{1}{T}dU + \frac{p}{T}dV ,$$

$$dU = \left(\frac{\partial U}{\partial T}\right)_V dT + \left(\frac{\partial U}{\partial V}\right)_T dV = C_V dT + \left(\frac{\partial U}{\partial V}\right)_T dV .$$

Insertion yields:

$$dS = \frac{1}{T}\left(\frac{\partial U}{\partial T}\right)_V dT + \frac{1}{T}\left[\left(\frac{\partial U}{\partial V}\right)_T + p\right]dV . \tag{2.58}$$

Since dS is a total differential, the integrability conditions (1.33) are fulfilled:

$$\frac{1}{T}\left[\frac{\partial}{\partial V}\left(\frac{\partial U}{\partial T}\right)_V\right]_T = -\frac{1}{T^2}\left[\left(\frac{\partial U}{\partial V}\right)_T + p\right]$$

$$+\frac{1}{T}\left\{\left[\frac{\partial}{\partial T}\left(\frac{\partial U}{\partial V}\right)_T\right]_V + \left(\frac{\partial p}{\partial T}\right)_V\right\} .$$

Since dU, too, is a total differential, this expression simplifies to:

$$\left(\frac{\partial U}{\partial V}\right)_T = T\left(\frac{\partial p}{\partial T}\right)_V - p . \tag{2.59}$$

The right-hand side is determined only by the equation of state. In case of a known heat capacity C_V the internal energy $U(T, V)$ can thus be derived exclusively from the equation of state.

Examples

1) Ideal gas

$$\left(\frac{\partial U}{\partial V}\right)_T = T\frac{nR}{V} - p = 0 \ . \tag{2.60}$$

The statement of the Gay-Lussac experiment that the internal energy of the ideal gas does not depend on the volume is therefore an immediate consequence of the basic relation of thermodynamics:

$$U = U(T) = C_V T + \text{const} \ . \tag{2.61}$$

2) Van der Waals gas

With the equation of state (1.14) inserted into (2.59) one finds (see Exercise 2.9.11):

$$\left(\frac{\partial U}{\partial V}\right)_T = a\frac{n^2}{V^2} \ . \tag{2.62}$$

Due to the particle interactions the internal energy is now volume-dependent:

$$U = U(T, V) = C_V T - a\frac{n^2}{V} + \text{const} \tag{2.63}$$

($C_V = $ const assumed!) **3) Photon gas**

When we insert (2.8) into (2.59) we get:

$$\varepsilon(T) = \frac{1}{3}T\frac{d\varepsilon}{dT} - \frac{1}{3}\varepsilon(T) \iff 4\varepsilon(T) = T\frac{d\varepsilon}{dT} \ .$$

The solution is the **Stefan-Boltzmann law**:

$$\varepsilon(T) = \text{const} \cdot T^4 \ . \tag{2.64}$$

As a consequence of the first law we had already found for the difference of the heat capacities C_p and C_V in (2.17)

$$C_p - C_V = \left[\left(\frac{\partial U}{\partial V}\right)_T + p\right]\left(\frac{\partial V}{\partial T}\right)_p \ .$$

That reads here with (2.59):

$$C_p - C_V = T\left(\frac{\partial p}{\partial T}\right)_V\left(\frac{\partial V}{\partial T}\right)_p \ . \tag{2.65}$$

We see that this difference is determined exclusively by the thermal equation of state.

The right-hand side can be expressed by relatively simply measurable **response functions**.

Definition 2.8.1

1. **Isobaric thermal expansion coefficient**

$$\beta = \frac{1}{V}\left(\frac{\partial V}{\partial T}\right)_p .$$

$$(2.66)$$

2. **Isothermal (adiabatic) compressibility**

$$\kappa_{T(S)} = -\frac{1}{V}\left(\frac{\partial V}{\partial p}\right)_{T(S)},$$

$$(2.67)$$

With the chain rule (Exercise 1.6.2)

$$\left(\frac{\partial p}{\partial T}\right)_V \left(\frac{\partial T}{\partial V}\right)_p \left(\frac{\partial V}{\partial p}\right)_T = -1$$

$$(2.68)$$

as well as

$$\left(\frac{\partial T}{\partial V}\right)_p = \frac{1}{\left(\frac{\partial V}{\partial T}\right)_p} = \frac{1}{V\beta}$$

it follows:

$$\left(\frac{\partial p}{\partial T}\right)_V = \frac{\beta}{\kappa_T} .$$

$$(2.69)$$

Inserted into (2.65) that yields:

$$C_p - C_V = \frac{T V \beta^2}{\kappa_T} .$$

$$(2.70)$$

Mechanical stability of the system requires

$$\kappa_T \geq 0 .$$

$$(2.71)$$

This pretty plausible relation can even explicitly be proven within the framework of Statistical Mechanics. It has the consequence:

$$C_p > C_V .$$

$$(2.72)$$

This relation is intuitively clear since at constant pressure p the same temperature increase dT needs '*more δQ*' than the process at constant volume. This is so because in the case of C_p volume work also has to be done. On the other hand, volume work does not appear for C_V because of $V = $ const, i.e., $dV = 0$.

So far we have presumed T and V to be the independent state variables. Experimental boundary conditions, however, could be different, so that T and p or V and p appear as *more conveniently measurable*. One has then to formulate the relevant state functions in the corresponding set of variables. This we want to demonstrate finally for the example of the entropy. We derive the so-called *T dS-* **equations**.

1. $S = S(T, V)$

   This is the case which we have already discussed. If one inserts (2.59) into (2.58) and exploits (2.69) then it remains:

$$T\,dS = C_V\,dT + T\frac{\beta}{\kappa_T}\,dV \ . \tag{2.73}$$

Even the calculation of the entropy requires, besides the thermal equation of state ($\Longrightarrow \beta, \kappa_T$), only the knowledge of C_V.

2. $S = S(T, p)$

$$V = V(T, p) \implies dV = \left(\frac{\partial V}{\partial T}\right)_p dT + \left(\frac{\partial V}{\partial p}\right)_T dp \ .$$

This is inserted into (2.58):

$$
\begin{aligned}
T\,dS &= \left(\frac{\partial U}{\partial T}\right)_V dT + \left[\left(\frac{\partial U}{\partial V}\right)_T + p\right]\left(\frac{\partial V}{\partial T}\right)_p dT \\
&\quad + \left[\left(\frac{\partial U}{\partial V}\right)_T + p\right]\left(\frac{\partial V}{\partial p}\right)_T dp \\
&\overset{(2.16)}{=} C_p\,dT + T\left(\frac{\partial p}{\partial T}\right)_V \left(\frac{\partial V}{\partial p}\right)_T dp \\
&\overset{(2.69)}{=} C_p\,dT + T\left(\frac{\beta}{\kappa_T}\right)(-V\kappa_T)\,dp \ .
\end{aligned}
$$

Therewith the $T\,dS$-equation in the variables (T, p) reads:

$$T\,dS = C_p\,dT - T\,V\,\beta\,dp \ . \tag{2.74}$$

3. $S = S(V, p)$

$$T = T(p, V) \implies dT = \left(\frac{\partial T}{\partial p}\right)_V dp + \left(\frac{\partial T}{\partial V}\right)_p dV \ .$$

Insertion into (2.58),

$$T\, dS = C_V\, dT + T \left(\frac{\partial p}{\partial T}\right)_V dV \ ,$$

yields the intermediate result:

$$T\, dS = C_V \left(\frac{\partial T}{\partial p}\right)_V dp + \left[C_V \left(\frac{\partial T}{\partial V}\right)_p + T\left(\frac{\partial p}{\partial T}\right)_V\right] dV \ . \qquad (2.75)$$

It follows with (2.69):

$$C_V \left(\frac{\partial T}{\partial p}\right)_V = C_V \frac{\kappa_T}{\beta} \ ,$$

$$\left[C_V \left(\frac{\partial T}{\partial V}\right)_p + T\left(\frac{\partial p}{\partial T}\right)_V\right] = \left(\frac{\partial T}{\partial V}\right)_p \left[C_V + T\left(\frac{\partial p}{\partial T}\right)_V \left(\frac{\partial V}{\partial T}\right)_p\right]$$

$$\overset{2.65}{=} C_p \left(\frac{\partial T}{\partial V}\right)_p \overset{2.66}{=} \frac{C_p}{V\beta} \ .$$

We have found therewith the third $T\, dS$-equation:

$$T\, dS = C_V \frac{\kappa_T}{\beta} dp + \frac{C_p}{V\beta} dV \ . \qquad (2.76)$$

If one evaluates these $T\, dS$-equations in particular for adiabatic-reversible processes ($S = \text{const}$) some additional useful relationships are found:

$$(2.73) \implies \left(\frac{\partial V}{\partial T}\right)_S = -\frac{C_V \kappa_T}{T \beta} \ ,$$

$$(2.74) \implies \left(\frac{\partial p}{\partial T}\right)_S = \frac{C_p}{T V \beta}$$

$$\implies \frac{C_p}{V C_V \kappa_T} = -\left(\frac{\partial p}{\partial T}\right)_S \left(\frac{\partial T}{\partial V}\right)_S = -\left(\frac{\partial p}{\partial V}\right)_S = \frac{1}{V \kappa_S} \ .$$

This yields:

$$\frac{C_p}{C_V} = \frac{\kappa_T}{\kappa_S} \,. \tag{2.77}$$

Because of (2.72) it is thus always $\kappa_T > \kappa_S$. If we combine this equation with (2.70) then we can still solve explicitly for C_p and C_V, respectively:

$$C_p - C_V = \frac{T V \beta^2}{\kappa_T} = C_p - \frac{\kappa_S}{\kappa_T} C_p$$

$$\Longrightarrow C_p = \frac{T V \beta^2}{\kappa_T - \kappa_S} \,, \tag{2.78}$$

$$C_V = \frac{T V \beta^2 \kappa_S}{\kappa_T (\kappa_T - \kappa_S)} \,. \tag{2.79}$$

Relations analogous to those derived here for the fluid system (gas-liquid) are also valid for **magnetic systems**, if one uses the proper *response functions*. The compressibility is replaced by the
isothermal (adiabatic) susceptibility

$$\chi_{T(S)} = \left(\frac{\partial M}{\partial H} \right)_{T(S)} = \frac{1}{V} \left(\frac{\partial m}{\partial H} \right)_{T(S)} \,. \tag{2.80}$$

But note that, in contrast to compressibilities, susceptibilities can become negative also. (Diamagnetism, cf. section 3.4.2, Vol. 3.). The expansion coefficient has its analog in the quantity

$$\beta_H = \left(\frac{\partial M}{\partial T} \right)_H = \frac{1}{V} \left(\frac{\partial m}{\partial T} \right)_H \,, \tag{2.81}$$

which in the field of magnetism does not carry a special name.—The volume V is to be considered for magnetic systems as a constant and therewith as an unimportant parameter. It is therefore not a state variable as in the fluid system. If one regards the following mapping:

Magnet $\longleftrightarrow$	Gas
$\mu_0 H$	p
m	$-V$
$\frac{V}{\mu_0} \chi_{T(S)}$	$V \kappa_{T(S)}$
$C_{H,m}$	$C_{p,V}$
$V \beta_H$	$-V \beta$,

then one finds with (2.70), (2.77) and (2.78):

$$\chi_T = \mu_0 V \frac{T \beta_H^2}{C_H - C_m} , \tag{2.82}$$

$$C_H = \mu_0 V \frac{T \beta_H^2}{\chi_T - \chi_S} , \tag{2.83}$$

$$\frac{C_H}{C_m} = \frac{\chi_T}{\chi_S} . \tag{2.84}$$

2.9 Exercises

Exercise 2.9.1

1. Show that δQ is not a total differential. You can use the first law of thermodynamics and the fact that, in contrast, dU represents such a total differential.
2. Find for the example of the ideal gas an *integrating factor* $\mu(T, V)$, which turns δQ into a total differential $dy = \mu(T, V)\, \delta Q$ depending

   a) only on T ($\mu = \mu(T)$),
   b) only on V ($\mu = \mu(V)$)

Exercise 2.9.2 Show that for an ideal gas along the curve

$$p V^n = \text{const} \qquad (n = \text{const})$$

the ratio of the heat supplied and the work performed is constant.

Exercise 2.9.3 The volume change of an ideal gas takes place according to

$$\frac{dp}{p} = a \frac{dV}{V}$$

where a is a given constant. Determine $p = p(V)$, $V = V(T)$ and the heat capacity $C_a = \left(\frac{\delta Q}{dT}\right)_a$. How should a be chosen so that the change of state proceeds as an isobaric, isochoric, isothermal, or adiabatic process?

Exercise 2.9.4

1. Derive the general form of the thermal equation of state for a system that fulfills the relation:

$$\left(\frac{\partial U}{\partial V}\right)_T = 0$$

2. A gas with constant particle number fulfills the relations:

$$p = \frac{1}{V} f(T) \; ; \; \left(\frac{\partial U}{\partial V} \right)_T = bp' \; (b = \text{const.})$$

Determine the function $f(T)$!

Exercise 2.9.5 Prove that an adiabatic and an isotherm never intersect twice!

Exercise 2.9.6 For not too low temperatures, the Curie law represents the equation of state of the ideal paramagnet.

1. Show that it holds for the heat capacities

$$C_m = \left(\frac{\partial U}{\partial T} \right)_m \; ; \quad C_H = \left(\frac{\partial U}{\partial T} \right)_H + \mu_0 \frac{V}{C} M^2$$

(C = Curie constant).

2. Derive for adiabatic changes of state the following relationship:

$$\left(\frac{\partial m}{\partial H} \right)_{ad} = \frac{C_m}{C_H} \frac{\mu_0 m - \left(\frac{\partial U}{\partial H} \right)_T}{\mu_0 H - \left(\frac{\partial U}{\partial m} \right)_T} \; .$$

Exercise 2.9.7 A thermally isolated cylinder contains inside a frictionlessly movable, thermally insulating wall. The two chambers are filled by ideal gases with initial data as given in Fig. 2.15. The gas in the **left** chamber is heated up until the gas in the **right** chamber has reached the pressure $p_r = 3p_0$.

1. What is the amount of heat that the gas on the right side has absorbed? What is the work that has been done by the *right* gas?
2. What are the final temperatures on the left side and on the right side, respectively?
3. How much heat has been taken by the *left* gas?

Exercise 2.9.8 One mole of an ideal diatomic gas is relaxed at the constant temperature of 293 K quasi-statically from an initial pressure of $2 \cdot 10 \, \text{N/m}^2$ to a final pressure of $1 \cdot 10 \, \text{N/m}^2$. Work is performed thereby via a movable piston.

1. How large is the applied work?
2. How much of heat quantity has to be flown into the gas?
3. How large is the work performed if the expansion happens adiabatically instead of isothermally?
4. How does the temperature change thereby?

Fig. 2.15 Two ideal gases in a thermally isolated cylinder, separated by a frictionlessly movable, thermally insulating wall

Fig. 2.16 Schematic arrangement for the Rüchhardt experiment

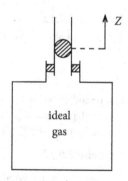

Exercise 2.9.9 A big vessel ends in a vertical tube with a smooth inner wall. The tube contains an easily movable but tightly fitting sphere (Fig. 2.16). The vessel is filled with an ideal gas.

The sphere is shifted a little bit out of its rest position and then released. It performs harmonic oscillations around the rest position (damping neglected!). The change of state that takes place can be considered to a good approximation to be adiabatic.

Calculate $\gamma = C_p/C_V$ as function of the period τ of the harmonic oscillation (Rüchhardt experiment).

Exercise 2.9.10 Two systems A and B, whose internal energies depend only on T, obey the equations of state

$$p V^2 = \alpha N T \quad (A) ,$$
$$p^2 V = \beta N T \quad (B) ,$$

where α, β represent constants with proper dimensions. Investigate whether for these systems an entropy is definable.

Exercise 2.9.11

1. Let the pressure p of a real gas be a linear function of temperature T:

$$p = \alpha(V) T + \beta(V) .$$

Then show that the heat capacity C_V cannot depend on volume V.
2. Calculate for the van der Waals gas the entropy $S = S(T, V)$ under the assumption that C_V does not depend on T.
3. Calculate the temperature change $\Delta T = T_2 - T_1$, which results due to the *free expansion* of a van der Waals gas ($C_V \neq C_V(T)$). Here *free expansion* means $U(T_1, V_1) = U(T_2, V_2)$.
4. For a reversible adiabatic change of state, calculate the *adiabatic equations* of the van der Waals gas.

Exercise 2.9.12 In simple approximation, for a solid, the following thermal equation of state is valid:

$$V = V_0 - \alpha p + \gamma T \ .$$

Let α and γ be material-specific parameters. Furthermore, let the heat capacity at constant pressure $C_p = $ const. be given. Calculate the heat capacity C_V and the internal energy $U(T, V)$ and $U(T, p)$, respectively!

Exercise 2.9.13

1. For the van der Waals gas, calculate the difference of the heat capacities $C_p - C_V$. For small model parameters a, b estimate the correction to the ideal gas!
2. What temperature change does the van der Waals gas experience with a quasistatic, reversible, adiabatic expansion from V_0 to $V > V_0$

Exercise 2.9.14 Given is a battery (reversible electro-chemical cell), which is ideal in the charge region q_a to $q_e > q_a$, i.e., the potential difference, which arises by the charge separation,

$$\varphi = \varphi(T, q) \equiv \varphi(T) \qquad q_a \le q \le q_e$$

is independent of the charge q. What is the heat ΔQ that must be given to the battery in connection with an isothermal charging ($q_a \to q_e$)? (work differential $\delta W = \varphi dq$)

Exercise 2.9.15 Given is a calorically ideal gas ($pV = nRT$, $C_V = $ const, $U = U(T)$).

1. Calculate its entropy $S = S(T, V)$.
2. Calculate the internal energy U as function of S and V.
3. Calculate the entropy change which appears due to a free expansion of the gas from V_1 to V_2.

Exercise 2.9.16 The equation of state of a real gas is given by the expression

$$pV = N k_B T \left(1 + \frac{N}{V} f(T)\right) ,$$

where $f(T)$ is an experimentally found function. Calculate the internal energy and the entropy of the gas with the precondition that

$$C_V = \frac{3}{2} N k_B - N k_B \frac{N}{V} \frac{d}{dt}\left(T^2 \frac{df}{dT}\right)$$

is valid.

Exercise 2.9.17 An ideal gas (n moles, C_V known) expands reversibly from volume V_1 to V_2,

1. at constant pressure p_0 (p_0 known!),
2. at constant temperature T_0 (T_0 known!),
3. adiabatically (initial pressure p_1 known!) .

Calculate the work ΔW, the exchanged heat ΔQ, and the entropy change ΔS as functions of V_1 and V_2.

Exercise 2.9.18 In an ideal gas the pressure is reversibly enhanced without any volume-change. Calculate ΔQ, ΔW and ΔS.

Exercise 2.9.19 The equation of state of a thermodynamic system (photon gas!) is

$$p = \alpha\, \varepsilon(T) ; \quad \alpha = \text{const} .$$

where $\varepsilon(T)$ is the internal energy per unit volume.

1. Determine the temperature dependence of the internal energy.
2. Calculate the entropy!

Exercise 2.9.20 Two different ideal gases with the mole numbers n_1 and n_2 are at first separated from one another in a vessel of the volume $V = V_1 + V_2$ by a heat-impermeable wall (Fig. 2.17). The pressure p on both sides is equal, and the temperatures are T_1 and T_2. The heat capacities of the two gases are identical.— Now the separating wall is removed.

1. What is the temperature after mixing?
2. What is the entropy change?
3. Demonstrate that the result in 2. cannot be correct for the case that the gases in the two chambers are identical and consist of **indistinguishable** particles (Gibb's paradox).

Exercise 2.9.21 A steel block of mass M with the constant heat capacity C_p is heated from an initial temperature T_a, which is equal to its ambient temperature, **on an isobaric way** to the temperature $T_0 > T_a$.

1. The heating may be due to a direct thermal contact of the block with a heat bath of the temperature T_0. What amount of heat is given by the bath to the block?

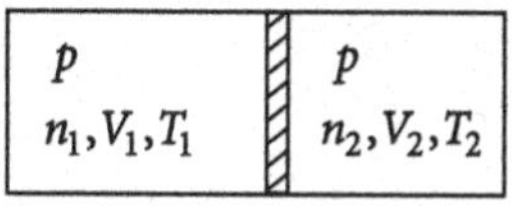

Fig. 2.17 Two different ideal gases with different temperatures, at first separated by a heat-impermeable wall

2. Between the steel block and the heat bath $HB(T_0)$ reversibly working Carnot machines are switched on, which in infinitesimal steps enhance the temperature of the block by corresponding heat transfers from the bath to the block. How much heat must be delivered in all by the bath in order to heat up the block (quasi-statically) to the temperature T_0?

3. Calculate the entropy changes of the heating processes 1) and 2), where one has to bear in mind that the heating of the block in 1) is irreversibly carried out. The heat supplies by the heat bath can be assumed to be reversible.

$$(\Delta S)_{HB}^{1),2)} = -\frac{\Delta Q_{1,2}}{T_0}$$

Exercise 2.9.22 A Carnot-thermodynamic cycle runs between the temperatures T_1 and T_2:

$$T_1 = 360\,\text{K}\,;\quad T_2 = 300\,\text{K}\,.$$

From the first heat bath the amount of heat removed is

$$\Delta Q_1 = 1\,\text{kJ}\,.$$

Calculate the executed work per cycle.

Exercise 2.9.23 An ideal gas with the heat capacity C_V runs reversibly through the cycle sketched in Fig. 2.18. p_a, V_a, T_a as well as p_b are known. Calculate

1. the volume V and the temperature T in the states b and c,
2. the exchanged amount of heat, the energy and entropy changes for each partial process,
3. the efficiency of the thermodynamic cycle.

Exercise 2.9.24 With an ideal gas the thermodynamic cycle sketched in Fig. 2.19 is reversibly performed. Calculate the efficiency as function of p_1 and p_2.

Exercise 2.9.25 Consider the reversible thermodynamic cycle of an ideal gas in the TS-plane, sketched in Fig. 2.20.

1. Calculate the heat quantities, which the system exchanges on the four segments, as functions of T_1, T_2 and S_1, S_2.

Fig. 2.18 Special reversible thermodynamic cycle for the ideal gas

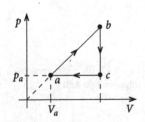

Fig. 2.19 Reversible
thermodynamic cycle for the
ideal gas consisting of
adiabatics and isobars

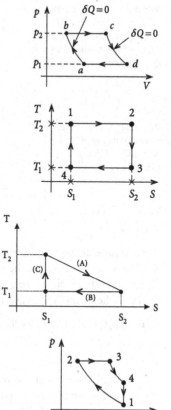

Fig. 2.20 Special
representation of the Carnot
process

Fig. 2.21 Reversible
thermodynamic cycle in the
TS-diagram

Fig. 2.22 The Diesel-process
as a special reversible
thermodynamic cycle for the
ideal gas

2. Determine the work done per cycle and calculate the efficiency η.
3. How does the pV-diagram of this process look like?

Exercise 2.9.26 An ideal gas runs reversibly through a thermodynamic cycle,
which consists of partial pieces (A), (B), and (C) as sketched in Fig. 2.21. Calculate
the various work outputs and heat exchange contributions! On which partial piece
does the system receive heat? Determine the efficiency η of the heat engine!

Exercise 2.9.27 Consider the reversible thermodynamic cycle for an ideal gas
sketched in Fig. 2.22 (*Diesel-process*). (1 → 2) and (3 → 4) are adiabatics. What
is the work performed by the system in one cycle? What is the amount of heat that
has to be taken in, what amount has to be given away?

Exercise 2.9.28 The thermodynamic cycle consisting of two adiabatic and two
isochoric branches, sketched in Fig. 2.23, is performed with an ideal gas as working
substance.

1. The diagram describes an idealized four-stroke combustion engine ('*Otto-motor*'). To which strokes do the various processes correspond?

Fig. 2.23 The combustion process in the Otto engine as an idealized thermodynamic cycle with two adiabatics and two isochores

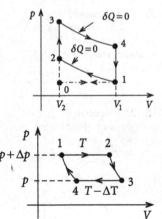

Fig. 2.24 Special reversible thermodynamic cycle for the derivation of the Clausius-Clapeyron equation

2. Calculate the work performed by the thermodynamic cycle!
3. How would you define the efficiency of the engine?
4. How is this efficiency related to that of a Carnot machine, which works between the highest and the lowest temperature?

Exercise 2.9.29 Consider the thermodynamic cycle (Carnot) sketched in Fig. 2.24.

- $1 \to 2$
  The liquid present at the point 1 in Fig. 2.24 with the volume V_1 evaporates at constant temperature T and constant pressure $p + \Delta p$. A part of the heat of evaporation is needed for overcoming the cohesive forces which are regained later with condensation. the rest serves for the expansion of the vapor ($V_1 \to V_2$).
- $2 \to 3$
  Adiabatic expansion with a cooling-down of ΔT.
- $3 \to 4$
  Isotherm compression where the vapor again completely condenses.
- $4 \to 1$
  Adiabatic compression with a temperature rise of ΔT.

Presuming that the volume changes on the adiabatics are negligibly small, derive by the use of the efficiency η of the Carnot cycle the Clausius-Clapeyron equation,

$$\frac{\Delta p}{\Delta T} = \frac{Q_D}{T (V_2 - V_1)} ,$$

which describes the curve of coexistence of gas and liquid (see Sect. 4.1.2).

Exercise 2.9.30 A certain amount of water is subject to a Carnot cycle between the temperatures $2\,°C$ and $6\,°C$. Because of the anomaly of the water it must absorb heat on both the isotherms. Is that a contradiction to Kelvin's formulation of the second law of thermodynamics?

Fig. 2.25 Thermodynamic Stirling cycle consisting of two isotherms and two isochores

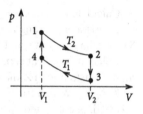

Exercise 2.9.31 An ideal gas runs through the thermodynamic Stirling cycle sketched in Fig. 2.25:

$$1 \to 2; \ 3 \to 4 : \quad \text{isothermal,}$$
$$2 \to 3; \ 4 \to 1 : \quad \text{isochoric.}$$

Calculate the efficiency!

Exercise 2.9.32 With an ideal gas a thermodynamic cycle is performed that consists of the following state changes:

$$(1)\{p_1, V_1\} \ \to \ (2)\{p_1, V_2\} \ \to \ (3)\{p_2, V_2\} \ \to \ (4)\{p_2, V_1\}$$
$$\to \ (1)\{p_1, V_1\} .$$

Thereby:

$$p_1 V_2 = p_2 V_1 .$$

1. Represent the process in the pV-plane and plot the isotherms.
2. Represent the process in the TV-plane and plot the isobars.
3. Represent the process in the pT-plane and plot the isochores.

Exercise 2.9.33 For a rubber thread the following connection between length L, tractive force Z and temperature T has been established:

$$L = L_0 + \frac{\alpha Z}{T} \qquad (L_0, \alpha : \text{constants}) .$$

The tractive force $Z = mg$ is realized by an connected weight of the mass m. For the heating up of the thread by the temperature difference $1\,\text{K}$ at constant length $L = L_0$ one needs, independent of the initial temperature, a constant amount of heat $C > 0$.

1. Show that the heat capacity of the thread at constant length L depends neither on the temperature T nor on the length L.

2. Calculate the internal energy $U(T, L)$ and the entropy $S(T, L)$. How do the adiabatic equations $T = T(L)$ and $Z = Z(L)$ look like?
3. Plot the isotherms and the adiabatics in a Z-L diagram.
4. Calculate the heat capacity C_Z at constant tractive force Z.
5. At constant load Z the thread shrinks when the temperature is raised from T_1 to $T_2 > T_1$. Which fractional amount β of the heat, introduced so that the weight is lifted, is converted into mechanical work?
6. The thermally isolated thread is stretched from L_1 to $L_2 > L_1$. Is the temperature then going to rise or drop?

Exercise 2.9.34 Consider once more the system from Exercise 2.9.33 and apply the partial results 1. to 3.

1. Sketch in the Z-L diagram a Carnot cycle diagram. In which direction must the cycle be traversed in order to operate as heat engine?
2. Let the two isotherms of the Carnot cycle belong to the temperatures T_1 and $T_2 > T_1$. ΔQ_1 and ΔQ_2 shall be the amounts of heat exchanged on these isotherms. Determine ΔQ_1, ΔQ_2 as well as the efficiency of the Carnot cycle.
3. Discuss a thermodynamic cycle which consists of only one isotherm and only one adiabatic where one of its corners lies at $L = L_0$.

Exercise 2.9.35 For a rubber thread it may hold, as in Exercise 2.9.33, the following connection between length L, tractive force Z and temperature T:

$$L = L_0 + \frac{\alpha Z}{T} .$$

The thread, at first freighted with Z, is abruptly released ($Z = 0$). Let the subsequent contraction of the thread be so rapid that thereby no heat exchange with the surroundings is possible. Calculate the entropy increase ΔS during this irreversible process as function of Z and T. How can the same final state be reached by a reversible process which can be used to calculate ΔS directly by integration of $\delta Q / T$?

Exercise 2.9.36 A crystal lattice, at certain lattice sites, has permanent magnetic moments. This system of moments is characterized by a magnetization of the form

$$M = \hat{C} \frac{H}{T} \qquad (Curie\ law,\ \hat{C} : Curie\ constant)$$

and by the following heat capacity at constant field H:

$$C_H^{(m)} = \hat{C} \mu_0 V \frac{H^2 + H_r^2}{T^2} \qquad (V, H_r : constants) ,$$

The crystal lattice has the heat capacity C_L, whose temperature dependence can be neglected in the following because of $C_L \gg C_H^{(m)}$. The total crystal is thermally isolated from surroundings.

1. Show that the amount of heat, received by the system of moments during a quasistatic process, is described by

$$\delta Q^{(m)} = C_H^{(m)} \, dT - \mu_0 V H \left(\frac{\partial M}{\partial H} \right)_T dH$$

The volume V is here an unimportant constant and not a thermodynamic variable!

2. Derive a condition equation for the temperature $T(H)$ of the magnetic system,

   a) if no heat exchange takes place between magnetic system and crystal lattice;
   b) if both partial systems are permanently in thermal equilibrium!

3. Let the total system have an initial temperature T^* and let it be in a field $H = H^*$.

   a) The field is switched off so rapidly that no heat exchange between moment system and crystal lattice takes place, but otherwise also so slowly that the process can be treated as quasistatic. Which temperature T_0 does the moment system have after switching off the field?
   b) Because of the subsequent heat exchange between the partial systems it comes to a thermal equilibrium with the temperature T_f. Calculate T_f!

4. Starting at the same initial state as in (3a) the field is switched off so slowly that both the partial systems are always in thermal equilibrium. Which final temperature $\widehat{T}_f$ is now reached?

5. Discuss the results from 3. and 4.

   a) Are the processes reversible?
   b) Why are the final temperatures T_f and $\widehat{T}_f$ not identical? Which temperature is higher?

2.10 Self-Examination Questions

To Sect. 2.1

1. What is the essential statement of the first law of thermodynamics?
2. How is the internal energy U defined? How does U change when the system has run through a thermodynamic cycle?
3. What is to be understood by the chemical potential μ?
4. Which relationship is called *caloric* and which is called *thermal* state equation?
5. Formulate the first law of thermodynamics for isolated, closed and open systems!

To Sect. 2.2

1. What is the general definition of a heat capacity? Which physical statements are delivered by it?
2. What is the difference between heat capacity, specific heat, and molecular heat?
3. Explain why for the ideal gas we must have $C_p > C_V$!

To Sect. 2.3

1. What does one understand by an adiabatic state change?
2. What are the three adiabatic equations of the ideal gas?
3. What can be said about the heat capacities C_V and C_p of the black body radiator?
4. Formulate the adiabatic equations of the black body radiator!
5. What is an isotherm?
6. In the pV-diagram for the ideal gas, plot qualitatively an isochore, isobar, isotherm, and an adiabatic. All the curves thereby should have a common point (p_0, V_0).

To Sect. 2.4

1. Why is the first law of thermodynamics not sufficient for the full description of thermodynamic processes?
2. What does one understand by a perpetuum mobile of the second kind?
3. What is the physical content of the second law of thermodynamics? Give Kelvin's as well as Clausius' version of the second law!
4. What is the definition of a *heat engine*?
5. What is meant by the *efficiency η* of a heat engine?

To Sect. 2.5

1. Define the Carnot cycle!
2. What is a *heat pump*?
3. What is the efficiency of the Carnot machine?
4. What can be said about the efficiency of an arbitrary, reversibly and periodically working machine?

To Sect. 2.6

1. Which universal property of the Carnot machine can be exploited to fix the absolute thermodynamic temperature scale?
2. For the proof of the universality of the efficiency of reversible thermodynamic cycles, how is the fact used that the working substance is an ideal gas?
3. Sketch the idea of how one can fix an absolute, substance independent temperature with the aid of reversible thermodynamic cycles!

To Sect. 2.7

1. What is the statement of the Clausius inequality?
2. How is the entropy S defined? Is it a unique definition?
3. Which integrating factor changes the differential form δQ into the total differential dS?
4. How does one determine the entropy if the changes of state take place irreversibly?
5. How do we formulate *mathematically* the second law of thermodynamics?
6. What do we understand by the basic relation of thermodynamics?
7. What would you denote as the central terms of the phenomenological thermodynamics?
8. How does the entropy of an isolated system behave during the time when processes are still running? What can be said about the entropy after the system has reached its equilibrium?
9. Irreversible processes are characterized by what?
10. Describe a reversible and an irreversible possibility to expand the ideal gas isothermally!

To Sect. 2.8

1. List a few important conclusions, which follow from the fact that dS and dU are total differentials!
2. Show that in the case of known heat capacity C_V the internal energy $U(T, V)$ can be derived from the equation of state alone!
3. Give reasons for the inequality $C_p > C_V$!
4. Show that the result of the Gay-Lussac experiment is a direct consequence of the basic relation of thermodynamics!
5. Verify the Stefan-Boltzmann law with the basic relation of thermodynamics for the photon gas!
6. What are denoted as the $T dS$-equations?
7. Which analogies exist between the fluid and the magnetic system?

Chapter 3
Thermodynamic Potentials

3.1 'Natural' State Variables

We know that reversible changes of state in actuality have to proceed quasi-statically as processes between equilibrium states. For such reversible processes, the

basic relation of thermodynamics

reads in its most general form:

$$dU = T\,dS + \sum_{i=1}^{m} F_i\,dq_i + \sum_{j=1}^{\alpha} \mu_j\,dN_j\,. \qquad (3.1)$$

Here we obviously have:

$$U = U(S, \mathbf{q}, \mathbf{N})\,. \qquad (3.2)$$

In particular, for **gases** we find with $\{\mathbf{F}, \mathbf{q}\} \rightarrow \{-p, V\}$:

$$dU = T\,dS - p\,dV + \sum_{j=1}^{\alpha} \mu_j\,dN_j\,, \qquad (3.3)$$

$$U = U(S, V, \mathbf{N})\,. \qquad (3.4)$$

Since dU is a total differential, one can consider the internal energy U also as the **generator** of the dependent variables. For the gas, e.g., one directly deduces from (3.3):

$$T = \left(\frac{\partial U}{\partial S}\right)_{V,\mathbf{N}} \;;\quad -p = \left(\frac{\partial U}{\partial V}\right)_{S,\mathbf{N}} \;;\quad \mu_j = \left(\frac{\partial U}{\partial N_j}\right)_{S,V,N_{i,\,i\neq j}}\,. \qquad (3.5)$$

© Springer International Publishing AG 2017
W. Nolting, *Theoretical Physics 5*, DOI 10.1007/978-3-319-47910-1_3

The experimentally important *response*-functions result from the second derivatives:

$$\left(\frac{\partial^2 U}{\partial S^2}\right)_{V,N} = \left(\frac{\partial T}{\partial S}\right)_{V,N} = \left[\left(\frac{\partial S}{\partial T}\right)_{V,N}\right]^{-1} \overset{\text{rev}}{=} \frac{T}{C_V}$$

$$\implies C_V = T \left[\left(\frac{\partial^2 U}{\partial S^2}\right)_{V,N}\right]^{-1} . \tag{3.6}$$

The second derivative of the internal energy with respect to the volume leads to the adiabatic compressibility:

$$\left(\frac{\partial^2 U}{\partial V^2}\right)_{S,N} = -\left(\frac{\partial p}{\partial V}\right)_{S,N} = \frac{1}{V \kappa_S}$$

$$\implies \kappa_S = \frac{1}{V}\left[\left(\frac{\partial^2 U}{\partial V^2}\right)_{S,N}\right]^{-1} . \tag{3.7}$$

Further useful relations result eventually from the fact that dU as a total differential has to fulfill the corresponding integrability conditions:

$$\left(\frac{\partial T}{\partial V}\right)_{S,N} = -\left(\frac{\partial p}{\partial S}\right)_{V,N} \quad ; \quad \left(\frac{\partial T}{\partial N_i}\right)_{V,S,N_{j,j\neq i}} = \left(\frac{\partial \mu_i}{\partial S}\right)_{V,N} . \tag{3.8}$$

These expressions are called **Maxwell relations**.

The equations (3.5) to (3.7) make clear that the full equilibrium behavior of the system, equations of states included, is uniquely fixed as soon as

$$U = U(S, \mathbf{q}, \mathbf{N})$$

is known. A quantity which is able to do that is called a

thermodynamic potential.

Its independent state variables are called:

natural variables

The natural variables of the **internal energy** are thus

$$\{S, \mathbf{q}, \mathbf{N}\}, \text{ especially for the } \mathbf{gas} \ \{S, V, \mathbf{N}\} .$$

The nomenclature *'potential'* is due to a formal analogy to the potential of Classical Mechanics. There one gets the components of the forces directly as first derivatives of the potential with respect to the coordinates.—One therefore speaks of the **natural variables** of a thermodynamic potential provided the corresponding dependent variables can be **directly** found by differentiating the potentials. According to (3.5), for the internal energy U, this is just the case when we represent U for the gas as a function of S, V and $\mathbf{N}$. These are the variables through which the *differential properties* of U are especially **simple and complete**. The caloric equation of state

$$U = U(T, V, \mathbf{N})$$

is therefore **not** a suitable thermodynamic potential. Because of

$$\left(\frac{\partial U}{\partial T}\right)_V = C_V \; ; \quad \left(\frac{\partial U}{\partial V}\right)_T = T\left(\frac{\partial p}{\partial T}\right)_V - p$$

the dependent state variables S and p do not follow directly as the first derivatives of U.

There are further aspects which mark the natural variables. We will later formulate equilibrium conditions for thermodynamic systems, in the following sense. In systems, in which the natural variables are kept constant, all possible irreversible processes proceed in such a way that the thermodynamic potential becomes extremal at equilibrium.

The introduction of other thermodynamic potentials, which we do in the next section, serves only the purpose to finding other *energy functions*, which are as functions of other sets of variables equally simple as U is as a function of $\{S, \mathbf{q}, \mathbf{N}\}$.

If one solves the basic relation of thermodynamics (3.1) for dS,

$$dS = \frac{1}{T}dU - \frac{1}{T}\sum_{i=1}^{m} F_i \, dq_i - \frac{1}{T}\sum_{j=1}^{\alpha} \mu_j \, dN_j \, , \tag{3.9}$$

then one realizes that $S = S(U, \mathbf{q}, \mathbf{N})$ also represents a thermodynamic potential.

3.2 Legendre Transformation

One disadvantage of using the internal energy U as thermodynamic potential is obvious. The natural variables are rather uncomfortable since, for instance, the entropy S is not easy to control. One introduces therefore, in view of the experimental boundary conditions, other thermodynamic potentials which utilize as natural variables just such quantities, which are more directly accessible for the experiment. The transition from one set of variables to another one is performed by applying the

Legendre transformation

which was introduced in Sect. 2.1 of Vol. 2 of this ground course on **Theoretical Physics**. We apply this technique to the internal energy U where we always intend to discuss the gas as a special example.

1.

$$\textbf{Free energy} \quad F = F(T, \mathbf{q}, \mathbf{N}) \,,$$
$$\textit{Gas:} \quad F = F(T, V, \mathbf{N}) \,.$$

The *former*, i.e. related to U, independent variable S is now replaced by the temperature T:

$$F = U - S \left(\frac{\partial U}{\partial S} \right)_{\mathbf{q}, \mathbf{N}} = U - TS \,. \tag{3.10}$$

The total differential dF results with (3.1) in:

$$dF = dU - d(TS) = dU - S\,dT - T\,dS \,,$$
$$dF = -S\,dT + \sum_{i=1}^{m} F_i\,dq_i + \sum_{j=1}^{\alpha} \mu_j\,dN_j \,. \tag{3.11}$$

This means in particular for the gas:

$$dF = -S\,dT - p\,dV + \sum_{j=1}^{\alpha} \mu_j\,d N_j \,. \tag{3.12}$$

The **natural variables** of the free energy thus are:

$$\{T, \mathbf{q}, \mathbf{N}\} \,; \quad \textit{Gas:} \ \{T, V, \mathbf{N}\} \,.$$

The dependent state quantities follow immediately from the first partial derivatives:

$$-S = \left(\frac{\partial F}{\partial T} \right)_{\mathbf{q}, \mathbf{N}} \,; \quad F_j = \left(\frac{\partial F}{\partial q_j} \right)_{T, \mathbf{N}, q_{i, i \neq j}} \,. \tag{3.13}$$

This means again in particular for the gas:

$$S = - \left(\frac{\partial F}{\partial T} \right)_{V, \mathbf{N}} \,; \quad p = - \left(\frac{\partial F}{\partial V} \right)_{T, \mathbf{N}} \,. \tag{3.14}$$

Furthermore, it holds, amongst others, the Maxwell relation:

$$\left(\frac{\partial S}{\partial V}\right)_{T,\mathbf{N}} = \left(\frac{\partial p}{\partial T}\right)_{V,\mathbf{N}} \quad \text{(Gas)} . \tag{3.15}$$

2.

$$\textbf{Enthalpy:} \quad H = H(S, \mathbf{F}, \mathbf{N}) ,$$
$$\textit{Gas:} \quad H = H(S, p, \mathbf{N}) .$$

Starting with U the generalized coordinates $\mathbf{q}$ are interchanged with the generalized forces $\mathbf{F}$:

$$H = U - \sum_{i=1}^{m} q_i \left(\frac{\partial U}{\partial q_i}\right)_{S,\mathbf{N},q_j \neq i} = U - \sum_{i=1}^{m} q_i F_i . \tag{3.16}$$

This means for the special case of a gas:

$$H = U + pV . \tag{3.17}$$

For the calculation of the total differential dH we utilize (3.1) also here:

$$dH = dU - \sum_{i=1}^{m} (dq_i F_i + q_i dF_i) \implies$$

$$dH = T\, dS - \sum_{i=1}^{m} q_i\, dF_i + \sum_{j=1}^{\alpha} \mu_j\, dN_j . \tag{3.18}$$

In the special case of the gas this reads:

$$dH = T\, dS + V\, dp + \sum_{j=1}^{\alpha} \mu_j\, dN_j . \tag{3.19}$$

The **natural variables** of the enthalpy are thus:

$$\{S, \mathbf{F}, \mathbf{N}\} ; \quad \textit{Gas:} \ \{S, p, \mathbf{N}\} .$$

Since H, too, represents a thermodynamic potential, the dependent state variables follow directly from the first partial derivatives:

$$T = \left(\frac{\partial H}{\partial S}\right)_{\mathbf{F},\mathbf{N}} ; \quad q_i = -\left(\frac{\partial H}{\partial F_i}\right)_{S,\mathbf{N},F_j \neq i} . \tag{3.20}$$

The second equation reads in the case of a gas:

$$V = \left(\frac{\partial H}{\partial p}\right)_{S,N} .$$

(3.21)

From (3.19) immediately follows the Maxwell relation:

$$\left(\frac{\partial T}{\partial p}\right)_{S,N} = \left(\frac{\partial V}{\partial S}\right)_{p,N} \quad (\text{Gas}) .$$

(3.22)

3.

Gibbs free enthalpy $\quad G = G(T, \mathbf{F}, \mathbf{N})$,

$Gas: \quad G = G(T, p, \mathbf{N})$.

Starting with U the variables S and $\mathbf{q}$ are now replaced by T and $\mathbf{F}$ using a corresponding Legendre transformation:

$$G = U - S\left(\frac{\partial U}{\partial S}\right)_{\mathbf{q},N} - \sum_{i=1}^{m} q_i \left(\frac{\partial U}{\partial q_i}\right)_{S,N,q_j \neq i} ,$$

$$G = U - TS - \sum_{i=1}^{m} q_i F_i .$$

(3.23)

For the gas we have:

$$G = U - TS + pV .$$

(3.24)

The total differential is again easily derivable:

$$dG = dU - T\,dS - S\,dT - \sum_{i=1}^{m} (q_i\,dF_i + F_i\,dq_i) .$$

When we insert (3.1),

$$dG = -S\,dT - \sum_{i=1}^{m} q_i\,dF_i + \sum_{j=1}^{\alpha} \mu_j\,dN_j ,$$

(3.25)

then we recognize that

$$\{T, \mathbf{F}, \mathbf{N}\} ; \quad Gas: \{T, p, \mathbf{N}\}$$

are the **natural variables** of the Gibbs free enthalpy. For the gas equation (3.25) takes the form

$$dG = -S\,dT + V\,dp + \sum_{j=1}^{\alpha} \mu_j\,dN_j \,.$$

(3.26)

The first partial derivatives of G with respect to the natural variables lead to the corresponding dependent state variables:

$$S = -\left(\frac{\partial G}{\partial T}\right)_{F,N} \; ; \quad q_i = -\left(\frac{\partial G}{\partial F_i}\right)_{T,N,F_j \neq i} \,.$$

(3.27)

For the gas the second equation reads:

$$V = \left(\frac{\partial G}{\partial p}\right)_{T,N} \,.$$

(3.28)

The Maxwell relation following from (3.26) is sometimes useful:

$$-\left(\frac{\partial S}{\partial p}\right)_{T,N} = \left(\frac{\partial V}{\partial T}\right)_{p,N} \,.$$

(3.29)

U, F, G and H are the most important thermodynamic potentials. A wealth of meaningful relationships result solely from the fact that dU, dF, dH and dG are total differentials.

3.3 Homogeneity Relations

It is an empirical fact that the internal energy U represents an

extensive state quantity.

This means that, if the extensive variables of the homogeneous phases of a thermodynamic system, in which the **intensive state variables** have the same values everywhere, are multiplied by a common factor, the internal energy U will correspondingly increase:

$$\left.\begin{array}{c} V \to \lambda V \\ N_j \to \lambda N_j \end{array}\right\} \quad \Longrightarrow \quad U \to \lambda U \,.$$

(3.30)

In the strict sense this is of course correct only when we have to take into consideration neither the interactions between the various partial systems nor surface effects. (In Statistical Mechanics we will introduce in this connection, and for this purpose, the **'thermodynamic limit'**!)

We want to now show that the other thermodynamic potentials, too, are extensive state quantities. Since

$$dU = T\,dS + \sum_{i=1}^{m} F_i\,dq_i + \sum_{j=1}^{\alpha} \mu_j\,dN_j$$

is extensive and the temperature T is by definition (Sect. 1.3) intensive, it must necessarily be:

$$S,\,dS \qquad \textbf{extensive,}$$

$$\mu_j \qquad \textbf{intensive,}$$

$$\sum_{i=1}^{m} F_i\,dq_i \qquad \textbf{extensive}$$

If the generalized coordinates q_i are chosen as extensive quantities then the corresponding forces F_i must be intensive and vice versa. By (3.10), (3.11), (3.16), (3.18), (3.23) and (3.25) one gets then immediately the assertion:

$$dF,\,dG,\,dH \quad \text{as well as} \quad F,\,G,\,H$$

extensive state quantities

Let us assume that the coordinates q_i are all extensive, as e.g. the volume V in the case of a gas, then one gets the important **homogeneity relations**:

$$F(T, \lambda\,\mathbf{q}, \lambda\,\mathbf{N}) = \lambda\,F(T, \mathbf{q}, \mathbf{N}) , \tag{3.31}$$

$$H(\lambda\,S, \mathbf{F}, \lambda\,\mathbf{N}) = \lambda\,H(S, \mathbf{F}, \mathbf{N}) , \tag{3.32}$$

$$G(T, \mathbf{F}, \lambda\,\mathbf{N}) = \lambda\,G(T, \mathbf{F}, \mathbf{N}) . \tag{3.33}$$

From the extensivity of G we derive an important conclusion. We differentiate both sides of the equation (3.33) with respect to λ and after the differentiation we put $\lambda = 1$:

$$G(T, \mathbf{F}, \mathbf{N}) = \frac{d}{d\lambda}\,G(T, \mathbf{F}, \lambda\,\mathbf{N})|_{\lambda=1}$$

$$= \left(\sum_{j=1}^{\alpha} \left(\frac{\partial G}{\partial (\lambda N_j)} \right)_{T, \mathbf{F}, N_{i \neq j}} N_j \right)_{\lambda=1} .$$

This yields the
Gibbs-Duhem relation

$$G(T, \mathbf{F}, \mathbf{N}) = \sum_{j=1}^{\alpha} \mu_j N_j , \qquad (3.34)$$

which looks especially simple if there is only one single type of particles ($\alpha = 1$):

$$G(T, \mathbf{F}, N) = \mu N . \qquad (3.35)$$

The chemical potential μ can thus be interpreted as the free enthalpy per particle.
Equation (3.34) can of course also be written as follows:

$$U - TS - \sum_{i=1}^{m} F_i q_i - \sum_{j=1}^{\alpha} \mu_j N_j = 0 . \qquad (3.36)$$

3.4   The Thermodynamic Potentials of the Ideal Gas

Before we derive, further generally valid properties of the thermodynamic potentials, we want to discuss some special applications. At first we calculate in this
section the potentials of the ideal gas explicitly. Let us assume that the gas consists
only of one particle type:

$$\{\mathbf{q}, \mathbf{F}, \mathbf{N}\} \;\rightarrow\; \{V, -p, N\} .$$

By $\overline{C}_{V,p}$ we denote the heat capacities per particle:

$$\overline{C}_{V,p} = \text{const} \qquad (3.37)$$

If the particle number N is constant, then it holds at first according to (2.58)
and (2.59):

$$dS = N\overline{C}_V \frac{dT}{T} + \frac{1}{T} p \, dV = N\overline{C}_V \, d\ln T + N k_B \, d\ln V .$$

This can be easily formally integrated:

$$S(T, V, N) = S(T_0, V_0, N) + N\overline{C}_V \ln \frac{T}{T_0} + N k_B \ln \frac{V}{V_0} . \qquad (3.38)$$

The entropy S is extensive and must therefore be homogeneous in the variables V and N:

$$S(T, \lambda V, \lambda N) \stackrel{!}{=} \lambda S(T, V, N) \qquad (\lambda \text{ real}) . \tag{3.39}$$

This is obviously not directly guaranteed by the intermediate result (3.38), in particular because of the $(\ln V)$-term on the right-hand side of the equation. We therefore have to place special requirements on $S(T_0, V_0, N)$:

$$S(T_0, V_0, \lambda N) + (\lambda N) \overline{C}_V \ln \frac{T}{T_0} + (\lambda N) k_B \ln \frac{\lambda V}{V_0} \stackrel{!}{=}$$

$$\stackrel{!}{=} \lambda S(T_0, V_0, N) + \lambda N \overline{C}_V \ln \frac{T}{T_0} + \lambda N k_B \ln \frac{V}{V_0} .$$

This is tantamount to

$$\lambda S(T_0, V_0, N) = S(T_0, V_0, \lambda N) + \lambda N k_B \ln \lambda .$$

Since λ can be chosen arbitrarily we are allowed to take especially $\lambda = N_0/N$:

$$S(T_0, V_0, N) = \frac{N}{N_0} S(T_0, V_0, N_0) + N k_B \ln \frac{N_0}{N} .$$

This we insert into (3.38):

$$S(T, V, N) = N \left\{ \sigma + \overline{C}_V \ln \frac{T}{T_0} + k_B \ln \frac{V/N}{V_0/N_0} \right\} . \tag{3.40}$$

σ here is now a real constant:

$$\sigma = \frac{1}{N_0} S(T_0, V_0, N_0) . \tag{3.41}$$

In the bracket there appear now, besides the constant σ, only intensive variables. S is thus homogeneous with respect to V and N.

In (3.40) we have represented the entropy actually not by its natural variables. These are according to (3.9) U, V, and N. By means of the caloric equation of state of the ideal gas,

$$U(T) = N \overline{C}_V T + \text{const} , \tag{3.42}$$

we can, however, easily replace in (3.40) T by U:

$$S(U, V, N) = N \left\{ \sigma + \overline{C}_V \ln \frac{U/N}{U_0/N_0} + k_B \ln \frac{V/N}{V_0/N_0} \right\} . \tag{3.43}$$

By solving for U we obtain the internal energy of the ideal gas as function of its natural variables S, V and N:

$$U = N \frac{U_0}{N_0} \exp\left[\frac{1}{\overline{C}_V}\left(\frac{1}{N}S - \sigma - k_B \ln \frac{V/N}{V_0/N_0}\right)\right].$$

With $k_B / \overline{C}_V = \gamma - 1$ we can also represent U in the following form:

$$U(S,V,N) = N\overline{C}_V T_0 \left(\frac{N_0 V}{N V_0}\right)^{1-\gamma} \exp\left[\frac{S}{N\overline{C}_V} - \frac{\sigma}{\overline{C}_V}\right] \tag{3.44}$$

The internal energy U of the ideal gas, if formulated as function of its **natural** variables, is apparently dependent on the volume. That is **not** a contradiction to the result of the Gay-Lussac experiment which refers to the caloric equation of state (3.42) and therefore to U in the variables T, V and N.

In the next step we calculate the free enthalpy, by using the Gibbs-Duhem relation (3.34). For this purpose we need the chemical potential μ, for which it applies according to (3.9):

$$-\frac{\mu}{T} = \left(\frac{\partial S}{\partial N}\right)_{U,V}. \tag{3.45}$$

We can calculate this explicitly with (3.43):

$$-\frac{\mu}{T} = \sigma + \overline{C}_V \ln \frac{U/N}{U_0/N_0} + k_B \ln \frac{V/N}{V_0/N_0}$$
$$+N\left[\overline{C}_V\left(-\frac{1}{N}\right) + k_B\left(-\frac{1}{N}\right)\right].$$

From this it follows with (3.42), if one sets the constant to zero:

$$\mu(T,V,N) = \left(k_B + \overline{C}_V - \sigma\right) T - \overline{C}_V T \ln \frac{T}{T_0} - k_B T \ln \frac{N_0 V}{N V_0}. \tag{3.46}$$

One realizes immediately that μ is an intensive variable:

$$\mu(T,V,N) \rightarrow \mu\left(T, \frac{V}{N}\right).$$

Accordingly we find

$$\mu(T,p,N) = \mu(T,p),$$

which can be derived from (3.46) with the aid of the equation of state and because of $\overline{C}_p = \overline{C}_V + k_B$:

$$\mu(T,p) = \left(\overline{C}_p - \sigma\right) T - \left(\overline{C}_p - k_B\right) T \ln \frac{T}{T_0} + k_B T \ln \frac{p/T}{p_0/T_0} . \tag{3.47}$$

With (3.34) it directly results the free enthalpy:

$$G(T,p,N) = N\mu(T,p). \tag{3.48}$$

For the free energy F we exploit:

$$F = G - pV = N\mu(T,V,N) - Nk_B T .$$

This with (3.46) leads to:

$$F(T,V,N) = N\left(\overline{C}_V - \sigma\right) T - N\overline{C}_V T \ln \frac{T}{T_0} - Nk_B T \ln \frac{V/N}{V_0/N_0} . \tag{3.49}$$

For the calculation of the enthalpy H it is advisable to start from

$$H = U + pV = N\left(\overline{C}_V + k_B\right) T = N\overline{C}_p T . \tag{3.50}$$

For the representation of H in its natural variables we have to find T as function of S, p and N. That succeeds by the use of (3.43) and the equation of state of the ideal gas:

$$S - N\sigma - Nk_B \left(\ln \frac{T}{T_0} - \ln \frac{p}{p_0} \right) = N\overline{C}_V \ln \frac{T}{T_0} .$$

With $\overline{C}_p = \overline{C}_V + k_B$ it follows further on:

$$\frac{S - N\sigma}{N\overline{C}_p} + \frac{\overline{C}_p - \overline{C}_V}{\overline{C}_p} \ln \frac{p}{p_0} = \ln \frac{T}{T_0} ,$$

$$T = T(S,p) = T_0 \left(\frac{p}{p_0}\right)^{(\gamma-1)/\gamma} \exp \left[\frac{S - N\sigma}{N\overline{C}_p} \right] . \tag{3.51}$$

The thermodynamic potentials of the ideal gas are therewith completely determined.

3.5 Entropy of Mixing

The considerations of the last section dealt with the potentials of an ideal gas which consists of a single particle type. In the case of multicomponent gases some additional considerations are necessary.

We consider two ideal gases consisting of two different types of particles:

a) Let the two gases be separated by a partition. In each chamber let there be the same pressure p and the same temperature T (Fig. 3.1). We then have the equations of state:

$$p V_1 = N_1 k_B T ,$$
$$p V_2 = N_2 k_B T .$$

Thermodynamic potentials are extensive. Therefore it holds for the internal energy U:

$$U_1 (T, N_1) + U_2 (T, N_2) = U(T) = \overline{C}_V (N_1 + N_2) T .$$

Assume that both types of particles have the same $\overline{C}_V$.

b) We now remove the partition. It starts an

irreversible intermixing

of the two gases which proceeds until it becomes a homogeneous mixture (Fig. 3.2). Except that nothing else happens! That is, there does not take place any performance of work performance or exchange of heat. The first law of thermodynamics then yields:

$$U = \text{const} = U(T) = \overline{C}_V (N_1 + N_2) T .$$

In particular the temperature remains constant. This means for the equation of state:

$$p (V_1 + V_2) = (N_1 + N_2) k_B T .$$

Fig. 3.1
Gedanken-experiment for the definition of the entropy of mixing

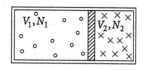

Fig. 3.2 Intermixing of two different ideal gases

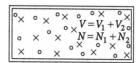

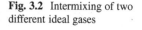

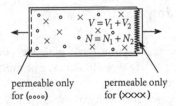

permeable only   permeable only
for (∘∘∘∘)   for (××××)

Fig. 3.3 Reversible auxiliary process for the intermixing of two different ideal gases (initial state with semipermeable walls)

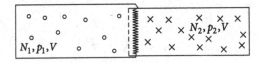

Fig. 3.4 Reversible auxiliary process for the mixing of two different ideal gases. Demixing is achieved by a semipermeable wall

$\{p, T, U\}$ do not change, but possibly the entropy S. About this we can state the following according to (2.53):

$$\Delta S = S(b) - S(a) \geq \int_a^b \frac{\delta Q}{T} = 0 \,. \tag{3.52}$$

Explicitly we can calculate ΔS only via a **reversible auxiliary process**:

$(b_1) = (b)$

The gas mixture is in two telescoped boxes, each of which is terminated at one side by a **semipermeable** wall. The left-hand side is **impermeable** for the particle type 2 while the particles of type 1 can diffuse through without any resistance. On the right-hand side it is exactly the opposite (Fig. 3.3).

We now pull apart the two boxes quasi-statically:

(b_2)

Thereby the gases are **reversibly** unmixed where each of them always retains its constant volume V (Fig. 3.4). The semipermeable walls move through the gas without resistance. The demixing does not need any work to be done:

$$\Delta W = 0 \,.$$

Fig. 3.5 Reversible auxiliary process for the demixing of two different ideal gases. After demixing by a semipermeable wall (Fig. 3.4) the system returns into the initial state by isothermal compression

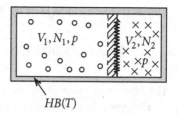

$HB(T)$

The temperature T does not change, i.e. $\Delta U = 0$, so that also $\Delta Q = 0$. The process is reversible, hence it holds:

$$\Delta S_{b_1 \to b_2} = 0 \ .$$

The pressures have changed:

$$p_1 V = N_1 \, k_B \, T \overset{(a)}{=} p \, V_1 \qquad\Longrightarrow\qquad p_i = p \, \frac{V_i}{V} \ ; \quad i = 1, 2$$
$$p_2 V = N_2 \, k_B \, T \overset{(a)}{=} p \, V_2 \qquad\qquad\qquad \text{(Dalton's law)} \ .$$

$\underline{(b_3) = (a)}$

By an isothermal, reversible compression, as described in Sect. 2.7, we take the system eventually back into the state (a). For this purpose we bring the total system into contact with a heat bath $HB(T)$ (Fig. 3.5).

Since the process is isothermal, no change of the internal energy takes place:

$$\Delta U = 0 \ .$$

However, work must be done on both the partial systems:

$$\Delta W = -\int_V^{V_1} p_1 \left(V' \right) dV' - \int_V^{V_2} p_2 \left(V' \right) dV'$$

$$= -N_1 \, k_B \, T \int_V^{V_1} \frac{dV'}{V'} - N_2 \, k_B \, T \int_V^{V_2} \frac{dV'}{V'}$$

$$= -k_B \, T \left\{ N_1 \ln \frac{V_1}{V} + N_2 \ln \frac{V_2}{V} \right\} \overset{!}{=} -\Delta Q_{\text{rev}} \ .$$

This corresponds to a change of entropy:

$$\Delta S_{b_2 \to b_3} = \frac{1}{T} \Delta Q_{\text{rev}} \ .$$

The total entropy change of $(b, 3) = (a)$ to $(b, 1) = (b)$ then amounts to:

$$\Delta S = S(b) - S(a) = k_B \left\{ N_1 \ln \frac{V}{V_1} + N_2 \ln \frac{V}{V_2} \right\} . \tag{3.53}$$

The entropy has thus increased!

The generalization from the case of two, here discussed, to α different gas species is straightforward:

$$\textbf{entropy of mixing:} \quad \Delta S = k_B \sum_{j=1}^{\alpha} N_j \ln \frac{V}{V_j} . \tag{3.54}$$

Bear in mind that the derivation of this expression presumes that the α ideal gases consist of pairwise distinguishable particles because otherwise the reversible auxiliary process would not work.

Since the entropy is an extensive quantity we can directly write it for the initial state of the gases **before** the mixing:

$$S_i = \sum_{j=1}^{\alpha} S\left(T, V_j, N_j\right) . \tag{3.55}$$

It only remains then to insert (3.38). The total entropy **after** the mixing is eventually calculated as follows:

$$
\begin{aligned}
S_f &= \Delta S + S_i \\
&= k_B \sum_{j=1}^{\alpha} N_j \ln \frac{V}{V_j} + \sum_{j=1}^{\alpha} N_j \left\{ \sigma + \overline{C}_V \ln \frac{T}{T_0} + k_B \ln \frac{V_j / N_j}{V_0 / N_0} \right\} \\
&= \sum_{j=1}^{\alpha} N_j \left\{ \sigma + \overline{C}_V \ln \frac{T}{T_0} + k_B \ln \frac{V / N_j}{V_0 / N_0} \right\} \\
\implies \quad S_f &= \sum_{j=1}^{\alpha} S\left(T, V, N_j\right) .
\end{aligned}
\tag{3.56}
$$

The entropies before and after the mixing thus differ only by the volumes which are filled by the various gas species. Before the mixing they are about the partial volumes V_j, after the mixing it is for all species the total volume V. For the derivation of the
entropy of mixing

$$\Delta S = \sum_{j=1}^{\alpha} \left\{ S\left(T, V, N_j\right) - S\left(T, V_j, N_j\right) \right\} \tag{3.57}$$

we could have applied from the beginning the formula (3.40), which is valid for all equilibrium states which are, at least in principle, achievable via a reversible process from a given reference point (index '0', constant σ (3.41)).

For gases with the same type of particles the expression (3.54) for the entropy of mixing seems to lead to a contradiction. Consider, for instance, two identical gases with

$$N_1 = N_2 = \frac{N}{2} \; ; \quad V_1 = V_2 = \frac{V}{2} \; .$$

Then (3.54) yields for the mixing of the two **identical** gases

$$\Delta S = N k_B \ln 2 \; , \tag{3.58}$$

although, of course, the entropy can **not** have changed (**Gibb's paradox**). This contradiction, however, does not exist in reality since (3.54) is not valid for identical gases. For such gases one has rather to argue as follows:

$$\left. \begin{array}{l} S_i = 2 S\left(T, \frac{V}{2}, \frac{N}{2}\right) \; , \\ S_f = S(T, V, N) = 2 S\left(T, \frac{V}{2}, \frac{N}{2}\right) \end{array} \right\} \quad \Longrightarrow \quad \Delta S = 0 \; .$$

One has therefore to act on the assumption that there is only one single gas species also after the mixing. For identical gas species, however, there do not exist *semipermeable* walls so that the above sketched *auxiliary process* for the demixing is not feasible.

3.6 Joule-Thomson Effect

We want to describe, as a further example for the application of thermodynamic potentials, the throttled adiabatic relaxation of a gas. The Joule-Thomson throttling experiment can be performed in such a way that a certain amount of gas of the initial volume V_1, with the initial temperature T_1 at a constant pressure p_1 is pressed through a porous wall into a room, the pressure of which p_2 is also kept constant. Let the final volume be V_2. One is interested in the change of the temperature of the gas from T_1 to T_2. The porous throttling zone is thought to prevent any arising of kinetic energy. The total system is thermally isolated (Fig. 3.6). According to the first law of thermodynamics at first we have:

$$\Delta U = U_2 - U_1 = \Delta W = -\int_{V_1}^{0} p_1 \, dV - \int_{0}^{V_2} p_2 \, dV = p_1 V_1 - p_2 V_2 \; .$$

Fig. 3.6 Schematic
arrangement for the
Joule-Thomson throttling
experiment

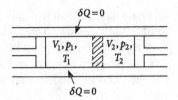

This means:

$$\Delta H = (U_2 + p_2 V_2) - (U_1 + p_1 V_1) = 0 . \tag{3.59}$$

The Joule-Thomson process is thus characterized by the fact that the enthalpy H
remains constant:

$$H = \text{const} \iff dH = T\,dS + V\,dp = 0 \quad (N = \text{const}) . \tag{3.60}$$

The
differential Joule-Thomson coefficient

$$\delta = \left(\frac{\partial T}{\partial p}\right)_H . \tag{3.61}$$

is of special interest. Since during the *throttling* we have $dp < 0$ ($p_2 < p_1$), it
follows

$$\delta > 0 : \quad \text{temperature lowering,}$$
$$\delta < 0 : \quad \text{temperature rise.}$$

We bring at first δ into a form which can be evaluated solely with the aid of the
equation of state of the gas:

$$S = S(T, p) \implies dS = \left(\frac{\partial S}{\partial p}\right)_T dp + \left(\frac{\partial S}{\partial T}\right)_p dT .$$

This we insert into (3.19) for dH ($N = \text{const} \iff dN = 0$):

$$dH = T\left(\frac{\partial S}{\partial T}\right)_p dT + \left[V + T\left(\frac{\partial S}{\partial p}\right)_T\right] dp . \tag{3.62}$$

By use of the Maxwell relation for the free enthalpy G,

$$\left(\frac{\partial S}{\partial p}\right)_T = -\left(\frac{\partial V}{\partial T}\right)_p , \tag{3.63}$$

we get further on:

$$dH = C_p \, dT + \left[V - T \left(\frac{\partial V}{\partial T} \right)_p \right] dp \ . \tag{3.64}$$

Therewith we have for the Joule-Thomson coefficient:

$$\delta = \left(\frac{\partial T}{\partial p} \right)_H = \frac{1}{C_p} \left[T \left(\frac{\partial V}{\partial T} \right)_p - V \right] \ . \tag{3.65}$$

Let us evaluate explicitly δ for two model systems:

1) Ideal gas

Using the equation of state one finds immediately:

$$T \left(\frac{\partial V}{\partial T} \right)_p = T \left(\frac{\partial}{\partial T} \frac{nRT}{p} \right)_p = \frac{nRT}{p} = V \ .$$

With the ideal gas one cannot achieve any cooling effect because of

$$\delta = 0 \ . \tag{3.66}$$

2) Van der Waals gas

We understand the left-hand side of the equation of state

$$\left(p + a \frac{n^2}{V^2} \right) (V - n b) = nRT$$

as an implicit function of T and p ($V = V(T, p)$) and differentiate for fixed pressure p with respect to T:

$$\left(-2 a \frac{n^2}{V^3} \right) (V - n b) \left(\frac{\partial V}{\partial T} \right)_p + \left(p + a \frac{n^2}{V^2} \right) \left(\frac{\partial V}{\partial T} \right)_p = nR \ .$$

From this it follows:

$$\left(\frac{\partial V}{\partial T} \right)_p \left(-a \frac{n^2}{V^2} + p + 2 a b \frac{n^3}{V^3} \right) = nR \ .$$

For the Joule-Thomson coefficient δ we need:

$$T\left(\frac{\partial V}{\partial T}\right)_p - V = \frac{nRT}{p - a\frac{n^2}{V^2} + 2ab\frac{n^3}{V^3}} - V . \qquad (3.67)$$

We are mainly interested in the so-called

inversion curve.

That is the curve in the pV-diagram, for which $\delta = 0$. According to (3.67) it must be fulfilled thereto

$$nRT = pV - a\frac{n^2}{V} + 2ab\frac{n^3}{V^2} .$$

With the van der Waals equation of state this is the case for

$$p_i = \frac{2a}{b}\frac{n}{V} - 3a\frac{n^2}{V^2} . \qquad (3.68)$$

The inversion curve $p_i(V)$ (Fig. 3.7) has obviously a zero at

$$V_0 = \frac{3}{2}bn$$

and at

$$V_{max} = 3nb$$

a maximum of the height $p_i^{max} = \frac{1}{3}a/b^2$.

Below the inversion curve $\delta > 0$, i.e., the adiabatic relaxation leads to a cooling of the van der Waals gas.

The Joule-Thomson process is irreversible and therefore involves an entropy production. It holds:

$$dS = \frac{1}{T}dH - \frac{V}{T}dp .$$

Fig. 3.7 Inversion curve of the van der Waals gas at the Joule-Thomson process

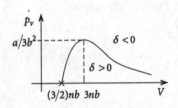

Because of $dH = 0$ and $dp < 0$, the entropy thus increases:

$$dS = -\frac{V}{T} dp > 0 .$$

3.7 Equilibrium Conditions

The thermodynamic potentials, as functions of their natural variables, are character-ized, in particular, by the fact that, by keeping certain variables constant and varying others, one can very easily recognize in which manner energy is exchanged with the surroundings. As an example, let us take the internal energy U of a gas:

$$dU = T\,dS - p\,dV \implies \begin{array}{ll} 1. & S = \text{const}: \quad dU = -p\,dV \quad (work) \\ 2. & V = \text{const}: \quad dU = T\,dS \quad (heat) \end{array}$$

The
basic relation of thermodynamics

$$T\,dS \geq \delta Q = dU - \sum_{i=1}^{m} F_i\,dq_i - \sum_{j=1}^{\alpha} \mu_j\,dN_j$$

can be brought into especially simple forms by the use of the thermodynamic potentials for the different contacts of the system and its surroundings. The potentials convey the possibility to describe the evolution of a thermodynamic system towards equilibrium and the equilibrium state itself. The various potentials are thereby suited to different experimental situations. Let us consider the most important special cases.

3.7.1 Isolated Systems

This situation we have already discussed in connection with the second law of thermodynamics (Sect. 2.7). Isolated systems are defined by:

$$dU = 0 \;(\delta Q = 0) ; \quad dq_i = 0 ; \quad dN_j = 0 . \tag{3.69}$$

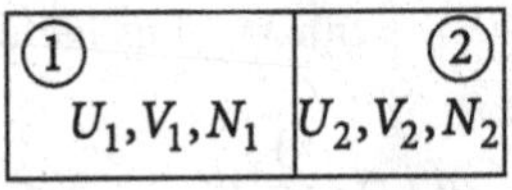

Fig. 3.8 An outwardly isolated system with an intermediate wall permeable to particles and energy

This means:

$$dS \geq 0 \, ,$$

$$dS = 0 \quad \text{in the equilibrium.} \tag{3.70}$$

So long as real irreversible processes run in isolated systems, these always behave such that the entropy thereby increases. The **entropy** is **maximal in the stationary equilibrium!**

As long as we take only the equilibrium value of the entropy $S = S(U, \mathbf{q}, \mathbf{N})$ as a basis, we cannot draw further conclusions beyond (3.70) since by assumption U, $\mathbf{q}$ and $\mathbf{N}$ are constant. We therefore create now in the form of a gedanken-experiment, a simple non-equilibrium situation from which further information can be derived.

Let the outwardly isolated system ($U = $ const, $V = $ const, $N_j = $ const) be decomposed by a wall into two parts. For simplicity, we assume that the gas in the chambers consists only of one single type of particles ($\alpha = 1$). The generalization to more than one type will pose no difficulty.

Let the wall be movable and permeable with respect to energy and particles!

V_1, V_2 as well as U_1, U_2 and N_1, N_2 are thus still variable, but with the **boundary conditions** (Fig. 3.8):

$$U = U_1 + U_2 = \text{const} \, ; \quad V = V_1 + V_2 = \text{const} \, ; \quad N = N_1 + N_2 = \text{const} \, .$$

The total entropy is additive:

$$S = S(U_1, V_1, N_1) + S(U_2, V_2, N_2) = S_1 + S_2 \, .$$

Since U, V, N are constant, we have:

$$dU_1 = -dU_2, \quad dV_1 = -dV_2, \quad dN_1 = -dN_2 \, .$$

The two partial systems will react until the equilibrium condition (3.70) is fulfilled:

$$0 = dS = dS_1 + dS_2$$

$$= \left\{ \left(\frac{\partial S_1}{\partial U_1} \right)_{V_1, N_1} - \left(\frac{\partial S_2}{\partial U_2} \right)_{V_2, N_2} \right\} dU_1$$

$$+ \left\{ \left(\frac{\partial S_1}{\partial V_1} \right)_{U_1, N_1} - \left(\frac{\partial S_2}{\partial V_2} \right)_{U_2, N_2} \right\} dV_1$$

$$+ \left\{ \left(\frac{\partial S_1}{\partial N_1} \right)_{U_1, V_1} - \left(\frac{\partial S_2}{\partial N_2} \right)_{U_2, V_2} \right\} dN_1$$

$$= \left(\frac{1}{T_1} - \frac{1}{T_2} \right) dU_1 + \left(\frac{p_1}{T_1} - \frac{p_2}{T_2} \right) dV_1$$

$$+ \left(-\frac{\mu_1}{T_1} + \frac{\mu_2}{T_2} \right) dN_1 \ . \tag{3.71}$$

U_1, V_1 and N_1 are independent state variables so that each of the brackets must separately vanish. The equilibrium is therefore characterized by

$$T_1 = T_2 = T ; \quad p_1 = p_2 = p ; \quad \mu_1 = \mu_2 = \mu \ . \tag{3.72}$$

We can proceed in the gedanken-experiment with the subdivision in order to come eventually *asymptotically* to the statement that

all points in an isolated system in equilibrium have the same temperature, the same pressure and the same chemical potential!

3.7.2 Closed System in a Heat Bath Without Exchange of Work

That means in detail:

$$\begin{aligned}
\textit{closed} \qquad &\Longrightarrow N_j = \text{const} \Longleftrightarrow dN_j = 0 \ , \\
\textit{in a heat bath} \qquad &\Longrightarrow T = \text{const} \Longleftrightarrow dT = 0 \ , \\
\textit{without exchange of work} &\Longrightarrow q_i \equiv \text{const} \Longleftrightarrow dq_i = 0 \ .
\end{aligned}$$

The basic relation then reads:

$$T \, dS \geq dU \ , T = \text{const} \quad \Longrightarrow \quad T \, dS = d(T S) \quad \Longrightarrow \quad d(U - T S) \leq 0 \ .$$

This leads to the important result:

$$dF \leq 0 \ ,$$

$$F = 0 \quad \text{in the equilibrium.} \tag{3.73}$$

For all irreversible processes which are possible under the given boundary conditions,
$$T = \text{const} , \quad \mathbf{q} = \text{const} , \quad \mathbf{N} = \text{const} ,$$
the free energy always decreases. F is minimal at equilibrium.

Fig. 3.9 System in a heat
bath with a dividing wall
permeable to particles

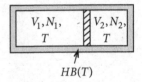

$$HB(T)$$

To get further statements we perform a similar gedanken-experiment as was done in
the last section for isolated systems, and that again for gas, as an example.

The system is subdivided by a wall which is freely movable and permeable to
particles. The heat bath ensures a constant temperature in both chambers (Fig. 3.9).

$$V = V_1 + V_2 = \text{const} \quad \Longrightarrow \quad dV_1 = -dV_2 \,,$$

$$N = N_1 + N_2 = \text{const} \quad \Longrightarrow \quad dN_1 = -dN_2 \,,$$

$$F = F(T, V_1, N_1) + F(T, V_2, N_2) = F_1 + F_2 \,.$$

At equilibrium we have:

$$0 = dF = dF_1 + dF_2$$

$$= \left\{ \left(\frac{\partial F_1}{\partial V_1} \right)_{N_1, T} - \left(\frac{\partial F_2}{\partial V_2} \right)_{N_2, T} \right\} dV_1$$

$$+ \left\{ \left(\frac{\partial F_1}{\partial N_1} \right)_{V_1, T} - \left(\frac{\partial F_2}{\partial N_2} \right)_{V_2, T} \right\} dN_1$$

$$= \{ -p_1 + p_2 \} \, dV_1 + \{ \mu_1 - \mu_2 \} \, dN_1 \,.$$

Since V_1 and N_1 are independent variables, it follows necessarily:

$$p_1 = p_2 = p \,; \quad \mu_1 = \mu_2 = \mu \,. \tag{3.74}$$

From this gedanken-experiment we can learn that *in a closed system (gas), whose
volume V is constant and which is in thermal contact with a heat bath, at
equilibrium, the pressure and the chemical potential have the same values at all
points in the system.*

3.7.3 Closed System in a Heat Bath with Constant Forces

That imposes the following preconditions:

$$dT = 0, \quad dN_j = 0, \quad dF_i = 0 \tag{3.75}$$

(Gas: $T = $ const, $N = $ const, $p = $ const). The basic relation now reads:

$$T\,dS = d(T\,S) \geq dU - \sum_{i=1}^{m} F_i\,dq_i = dU - d\left(\sum_{i=1}^{m} F_i\,q_i\right)$$

$$\Longrightarrow \quad d\left(U - \sum_{i=1}^{m} F_i\,q_i - T\,S\right) \leq 0.$$

This means:

$$dG \leq 0,$$

$$dG = 0 \quad \text{in the equilibrium.} \tag{3.76}$$

The free enthalpy (Gibb's potential) G always decreases during all irreversible processes which take place under the above boundary conditions. At equilibrium G is minimal!

The same gedanken-experiment as in the last section with the additional precondition $p = $ const in each chamber will now be performed with several types of particles (Fig. 3.10). The boundary conditions

$$N_j^{(1)} + N_j^{(2)} = N_j = \text{const} \quad \forall j \quad \Longleftrightarrow \quad dN_j^{(1)} = -dN_j^{(2)}$$

lead with

$$G = G\left(T, p, \mathbf{N}^{(1)}\right) + G\left(T, p, \mathbf{N}^{(2)}\right) = G_1 + G_2$$

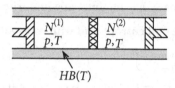

HB(T)

Fig. 3.10 System in a heat bath of the temperature T with a dividing wall permeable to particles, being bordered to the right and to the left by movable pistons which take care for a constant pressure p in the chambers

to the following expression:

$$dG = \sum_{j=1}^{\alpha} \left[\left(\frac{\partial G_1}{\partial N_j^{(1)}} \right)_{T, \mathbf{F}, N_{i, i \neq j}^{(1)}} - \left(\frac{\partial G_2}{\partial N_j^{(2)}} \right)_{T, \mathbf{F}, N_{i, i \neq j}^{(2)}} \right] dN_j^{(1)}$$

$$= \sum_{j=1}^{\alpha} \left\{ \mu_j^{(1)} - \mu_j^{(2)} \right\} dN_j^{(1)} \overset{!}{=} 0 .$$

This means that at equilibrium the chemical potential

$$\mu_j^{(1)} = \mu_j^{(2)} = \mu_j \tag{3.77}$$

assumes the same value everywhere in the system.

3.7.4 *Extremal Properties of* U *and* H

The equilibrium conditions derived so far are the in practice important ones. It is of course possible also to derive formally the conditions for U and H, which, however, are rather *unwieldy* because of the precondition to keep the entropy S constant.

1) Closed system of constant entropy without performance of work

This means:

$$dN_j = 0 , \quad dS = 0 , \quad dq_i = 0 \quad (\delta W = 0) . \tag{3.78}$$

The basic relation in this case yields:

$$dU \leq 0,$$
$$U = 0 \quad \text{in the equilibrium.} \tag{3.79}$$

For all processes, which can take place under the conditions (3.78), the **internal energy** U will never increase. It is **minimal at equilibrium**.

2) Closed system of constant entropy and with constant forces

With the boundary conditions

$$dN_j = 0 , \quad dS = 0 , \quad dF_i = 0 \tag{3.80}$$

the basic relation reads:

$$0 \geq dU - \sum_{i=1}^{m} F_i \, dq_i = d\left(U - \sum_{i=1}^{m} F_i \, q_i\right).$$

The consequence is:

$$dH \leq 0,$$

$$dH = 0 \quad \text{in the equilibrium.} \tag{3.81}$$

H decreases under the boundary conditions (3.80) for all processes, which can take place, and is **minimal at equilibrium!**

3.8   The Third Law of Thermodynamics (Nernst's Heat Theorem)

With the aid of the second law of thermodynamics we have introduced in Sect. 2.7 the fundamental state quantity **entropy**, which we could define, however, only except for an additive constant. Therefore, only entropy-**differences** between two points of the state space are unique, provided that they can be connected by a reversible state change. This, however, can not at all be taken for granted.

The equation of state, e.g. $f(T, p, V) = 0$ for a gas, defines a state plane in the (p, V, T)-space. If both the states A and B are on the same connected *sheet* of the state plane, then they can always be linked by a reversible path (Fig. 3.11). If, however, the state plane consists of two or more unconnected sheets (metastable phases, mixture of different substances, or something like that) then it is possible that such a reversible path does not exist. Then the undetermined constant prevents the direct comparison of the entropies in the states A and B.

The theorem of Nernst makes a statement about the behavior of the entropy for $T \to 0$ and abolishes therewith, at least partially, the mentioned indefiniteness. The theorem must be considered here as an empirical matter of fact, which can theoretically be reasoned only in the framework of Statistical Mechanics (Vol. 8).

Assertion 3.8.1 (Third Law of Thermodynamics)   The entropy of a thermodynamic system is at $T = 0$ a universal constant which can be chosen to be zero. That

Fig. 3.11 Path from state A to state B on a connected sheet of the state plane $f(T, p, V) = 0$

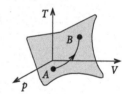

holds **independently** of the values of the other state variables:

$$\lim_{T \to 0} S(T, \mathbf{q}, \mathbf{N}) = 0 \,, \tag{3.82}$$

$$\lim_{T \to 0} S(T, \mathbf{F}, \mathbf{N}) = 0 \,. \tag{3.83}$$

This theorem is valid for **every** system and makes the entropy unique for **every** state. Out of this theorem, let us draw some experimentally verifiable **conclusions**.

1) Heat capacities

Assertion:

$$\lim_{T \to 0} C_{\mathbf{q}} = \lim_{T \to 0} T \left(\frac{\partial S}{\partial T} \right)_{\mathbf{q}} = 0 \,, \tag{3.84}$$

$$\lim_{T \to 0} C_{\mathbf{F}} = \lim_{T \to 0} T \left(\frac{\partial S}{\partial T} \right)_{\mathbf{F}} = 0 \,. \tag{3.85}$$

The heat capacities of all substances vanish at the absolute zero. This is experimentally uniquely confirmed. However, the *ideal gas* provides a contradiction since $C_V = $ const, $C_p = $ const. But on the other hand, the ideal gas is for $T \to 0$ surely not a realistic physical system (condensation!).

Proof Heat capacities are non-negative. It is therefore:

$$T \left(\frac{\partial S}{\partial T} \right)_{\ldots} = \left(\frac{\partial S}{\partial \ln T} \right)_{\ldots} \geq 0 \,.$$

We choose: $x = \ln T$. Then $T \to 0$ means nothing else but $x \to -\infty$. If it were

$$\lim_{x \to -\infty} \left(\frac{\partial S}{\partial x} \right)_{\ldots} = \alpha > 0 \,,$$

then, since as state quantity S is continuous, it would exist an x_0 with

$$-\infty < x \leq x_0 \quad \text{and} \quad \left(\frac{\partial S}{\partial x} \right) \geq \frac{\alpha}{2} > 0 \,.$$

This is tantamount to:

$$S(x_0) - S(x) = \int_x^{x_0} \left(\frac{\partial S}{\partial x'} \right) dx' \geq \frac{\alpha}{2} (x_0 - x) \,.$$

That would mean

$$S(x) \leq \frac{\alpha}{2} x + \text{const}$$

and would then, because of

$$\lim_{x \to -\infty} S(x) = -\infty,$$

contradict the Nernst theorem. The assumption $\alpha > 0$ thus must be wrong. It rather holds $\alpha = 0$, in compliance with the assertion.

For the **heat capacities of the gas** C_p, C_V it can be shown that the difference between them vanishes for $T \to 0$ even faster than T:

Assertion:

$$\lim_{T \to 0} \frac{C_p - C_V}{T} = 0. \tag{3.86}$$

Proof According to (2.65) it is:

$$\frac{C_p - C_V}{T} = \left(\frac{\partial p}{\partial T}\right)_V \left(\frac{\partial V}{\partial T}\right)_p.$$

We utilize the Maxwell relation of the free energy:

$$\left(\frac{\partial p}{\partial T}\right)_V = \left(\frac{\partial S}{\partial V}\right)_T.$$

The entropy S is at $T = 0$ independent of other variables, therefore it must be

$$\lim_{T \to 0} \left(\frac{\partial S}{\partial V}\right)_T = 0,$$

which proves the assertion.

2) Expansion coefficient

There are some other *response*-functions also, for which we can derive from the third law of thermodynamics statements about their $T \to 0$-behavior.

Assertion:

$$\beta = \frac{1}{V} \left(\frac{\partial V}{\partial T}\right)_p \xrightarrow[T \to 0]{} 0. \tag{3.87}$$

Proof We use one of the Maxwell relations of the free enthalpy:

$$\left(\frac{\partial V}{\partial T}\right)_p = -\left(\frac{\partial S}{\partial p}\right)_T .$$

By the same reasoning as above one finds:

$$\lim_{T \to 0}\left(\frac{\partial S}{\partial p}\right)_T = 0 .$$

That immediately yields the assertion.

The certainly most important implication of the third law concerns the

3) Unattainability of the absolute zero

One gets low temperatures by connecting in series adiabatic and isothermal processes with a suitable working substance, e.g. with a gas (*Linde method*) or with a paramagnet (*adiabatic demagnetization*). We sketch shortly the **principle:**

The entropy must depend, besides the temperature, on some other parameters x, e.g. on the pressure p for the gas, ($p_2 > p_1$), or on the magnetic field H for the paramagnet, ($H_2 > H_1$). One then performs the following process (Fig. 3.12):

$A \to B$

Entropy reduction by an isothermal change of the parameter x from x_1 to x_2. A certain heat amount has to be removed thereby, which in the reversible case is equal to $T_i \, \Delta S$.

$B \to C$

The system is now made thermally isolated and the parameter x is then brought back along an isentrope to its original value x_1. Thereby the temperature drops from T_i to T_f.

If the entropy curves would approach for $T \to 0$ different limiting values for different values of the parameter x, as sketched in Figs. 3.12, 3.13, then the absolute zero could be reached without any difficulty. Such an 'S-behavior' would, however, contradict the third law of thermodynamics, according to which all entropy curves should, for $T \to 0$, run into the origin. Figure 3.13 makes immediately clear that

Fig. 3.12 Entropy as function of temperature for two different parameters $(V, p, \ldots)$ with different limiting values for $T = 0$

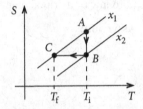

Fig. 3.13 Entropy as function of temperature for two different parameters $(V, p, \ldots)$ with different limiting values for $T = 0$ (*above*), a situation that would allow to reach the absolute zero. Unattainability of the absolute zero in the case of equal limiting values (*below*), corresponding to the third law

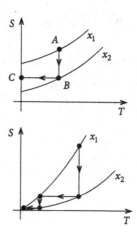

the point $T = 0$ can then be reached only asymptotically by an infinite number of partial steps. One can come arbitrarily close to it, but one never reaches it exactly.

We get the same result more formally by the following consideration:

Let us consider an adiabatic process,

$$S(T_1, x_1) \xrightarrow{\delta Q = 0} S(T_2, x_2) \ ,$$

for which it must hold according to the second law of thermodynamics ($T \, dS \geq \delta Q = 0$)

$$S(T_2, x_2) \geq S(T_1, x_1) \ ,$$

where the equality sign refers to a reversible transition. From the third law it follows now:

$$S(T_1, x_1) = \int_0^{T_1} \frac{C_x(x_1, T)}{T} \, dT \ ,$$

$$S(T_2, x_2) = \int_0^{T_2} \frac{C_x(x_2, T)}{T} \, dT \ .$$

$T_2 = 0$ would mean $S(T_2 = 0, x_2) = 0$ and therewith

$$\int_0^{T_1} \frac{C_x(x_1, T)}{T} \, dT \leq 0 \ .$$

This, however, is impossible because of $T_1 > 0$ and $C_x(x_1, T \neq 0) > 0$. T_2 can thus not be zero, what proves the unattainability of the absolute zero.

3.9 Exercises

Exercise 3.9.1 Let a system have the following properties:

a. The work executed by the system with the expansion from V_0 to V at constant temperature T_0 is

$$\Delta W_{T_0} = R T_0 \ln \frac{V}{V_0} .$$

b. Its entropy is

$$S = R \frac{V_0}{V} \left(\frac{T}{T_0} \right)^a ,$$

where V_0, T_0 and a ($a \neq -1$) are given constants.

Determine

1. the free energy,
2. the equation of state,
3. the work done at an **arbitrary** constant temperature T by expansion from V_0 to V.

Exercise 3.9.2 For the photon gas (black box radiator) we have:

$$U(T, V) = V \varepsilon (T) ; \quad p = \frac{1}{3} \varepsilon (T) .$$

Calculate therewith the thermodynamic potentials as functions of their natural variables.

Exercise 3.9.3 A spiral spring fulfills Hooke's law, i.e., the elongation x is proportional to the force $F_k = -k x$. The coefficient k depends on the temperature according to

$$k(T) = \frac{a}{T} ; \quad a > 0 .$$

How does the internal energy U of the system change, when the spiral spring is stretched up to x at constant temperature?

Exercise 3.9.4 A rubber band is elongated by an external force to the length L. It has the tension σ and the temperature T. At fixed length L one measures:

$$\sigma = \alpha T ; \quad \alpha > 0 .$$

1. Show that the internal energy depends only on temperature.
2. How does the entropy change for an isothermal lengthening of the band?
3. How does the temperature change if the band is elongated adiabatically?

Exercise 3.9.5 A paramagnetic material (heat capacity C_H known!) fulfills the Curie law. For a reversible adiabatic state change calculate the differential quotient:

$$\left(\frac{\partial T}{\partial H}\right)_S .$$

Exercise 3.9.6 Consider a rod of the length L with the thermal equation of state

$$Q = Q(T, L) .$$

Q is thereby the generalized force belonging to L, so that $\delta W = Q \, dL$ represents the corresponding work. The heat capacity $C_L(T, L_0)$ at fixed length L_0 is be known.

1. Calculate the heat capacity $C_L(T, L)$, the internal energy $U(T, L)$, the entropy $S(T, L)$, and the free energy $F(T, L)$ as functionals of $Q(T, L)$ and $C_L(T, L_0)$.
2. Evaluate the results of 1.) in particular for

$$Q(T, L) = aT^2(L - L_0) \; ; \; C_L(T, L_0) = bT ,$$

where a, b, L_0 are constants.
3. Calculate with the ansatz from 2.) the thermal expansion coefficient

$$\alpha = \frac{1}{L}\left(\frac{\partial L}{\partial T}\right)_Q .$$

4. Let the state of the rod change adiabatically reversibly from (T_1, L) to (T_2, L_0). Calculate T_2 as function of T_1, L and L_0!

Exercise 3.9.7 Let the free energy F of N identical particles in the volume V be given by:

$$F(T, V) = -N k_B T \ln C_0 V - N k_B T \ln C_1 (k_B T)^\alpha ,$$

C_0, C_1 : given constants > 0; α : given constant > 1.
Calculate

1. the entropy $S = S(T, V)$,
2. the pressure p,
3. the caloric equation of state $U = U(T, V)$,
4. the heat capacity C_V,
5. the isothermal compressibility κ_T.

Exercise 3.9.8 A closed volume V is divided by a wall into two sub-chambers with volumes V_1 and V_2. Each chamber contains N particles of the same type of a monoatomic ideal gas. Let the pressure be the same in both the chambers $p_1 = p_2 = p_0$, which can be realized by different temperatures T_1, T_2. The wall is now removed. Calculate the entropy of mixing, by the use of the entropy representation (3.43), as a function of T_1, T_2 and N. What happens for $T_1 = T_2$? Compare the result to that of Exercise 2.9.20!

Exercise 3.9.9 A paramagnetic substance has the isothermal magnetic susceptibility χ_T.

1. Calculate the magnetization-dependence of the free energy.
2. Derive with that the corresponding dependencies of the internal energy and the entropy.

Exercise 3.9.10 Calculate the free energy and the free enthalpy for a magnetic substance that obeys the Curie-Weiss law. Show at first that the heat capacity C_m depends only on temperature, and afterwards treat $C_m(T)$ as known.

Exercise 3.9.11 The volume of a system is be given as function of temperature and entropy, $V = V(T, S)$. Calculate the partial derivative of the enthalpy H with respect to the pressure p at constant volume.

Exercise 3.9.12 Consider a taut wire of piezoelectric material. *Piezoelectricity* means, with an isothermal or adiabatic change of the (mechanical) tension τ, a change of the electric polarization P is observed, or, with a change of the electric field strength E, a change of the length L and the tension τ results. (electrical (mechanical) work: $\delta W_e = VE\,dP$ ($\delta W_m = \tau\,dL$); V is not a variable).

1. Verify:

$$V\left(\frac{\partial P}{\partial \tau}\right)_{T,E} = \left(\frac{\partial L}{\partial E}\right)_{T,\tau}$$

    (V: volume, considered as constant).
2. How many different thermodynamic potentials exist for such a system?
3. How many integrability conditions are there?

Exercise 3.9.13 For a system with the particle number N, the internal energy U, the temperature T, the volume V, and the chemical potential μ show that the following relations are valid:

1. $\left(\dfrac{\partial U}{\partial N}\right)_{T,V} - \mu = -T\left(\dfrac{\partial \mu}{\partial T}\right)_{V,N}$,

2. $\left(\dfrac{\partial N}{\partial T}\right)_{V,\mu/T} = \dfrac{1}{T}\left(\dfrac{\partial N}{\partial \mu}\right)_{T,V}\left(\dfrac{\partial U}{\partial N}\right)_{T,V}$,

3. $\left(\dfrac{\partial U}{\partial T}\right)_{V,\mu/T} - \left(\dfrac{\partial U}{\partial T}\right)_{V,N} = \dfrac{1}{T}\left(\dfrac{\partial N}{\partial \mu}\right)_{T,V}\left(\dfrac{\partial U}{\partial N}\right)_{T,V}^2$.

Exercise 3.9.14 An ideal paramagnetic gas is characterized by the equations of state:

$$pV = Nk_BT \; ; \quad M = \frac{\alpha}{T} B_0 = \frac{m}{V} \; .$$

V is the volume, p the pressure, B_0 the magnetic induction, m the magnetic moment, M the magnetization, and T the temperature of the gas. α is a constant specific for a given material. Let the heat capacity be given by

$$C_{V,m} = \frac{3}{2} Nk_B \; .$$

1. Find the differentials of the internal energy $U = U(S, V, m)$ and the free energy $F = F(T, V, m)$!
2. Calculate by use of suitable integrability conditions the following differential quotients and evaluate explicitly the following for the ideal paramagnetic gas!

   a)
   $$\left(\frac{\partial S}{\partial V} \right)_{T,m}$$

   b)
   $$\left(\frac{\partial S}{\partial m} \right)_{T,V}$$

   c)
   $$\left(\frac{\partial U}{\partial V} \right)_{T,m}$$

   d)
   $$\left(\frac{\partial U}{\partial m} \right)_{T,V} \; .$$

3. Calculate the entropy $S(T, V, m)$ and the internal energy $U(T, V, m)$!
4. Show that the entropy fulfills the homogeneity relation!

Exercise 3.9.15 A paramagnetic substance has the isothermal susceptibility χ_T:

$$\chi_T = \left(\frac{\partial M}{\partial H} \right)_T = \frac{\mu_0}{V} \left(\frac{\partial m}{\partial B_0} \right)_T \; ; \quad (V = \text{const}, B_0 = \mu_0 H) \; .$$

The free energy F, the internal energy U, and the entropy S have already been calculated in Exercise 3.9.9 as functions of T and m.

1. How do these results read for magnetic systems with Curie-Weiss-behavior,

$$M = \frac{C}{T - T_C} H \quad \left(M = \frac{m}{V}; \quad C: \text{ Curie constant (1.26)} \right),$$

   i.e. for ferromagnets at temperatures $T > T_C$?
2. For the substance from 1. if in addition we have:

$$C_m(T, m = 0) = \gamma T \quad (\gamma > 0).$$

   Calculate therewith $F(T, m)$, $S(T, m)$, $S(T, H)$ as well as $U(T, m)$!
3. Calculate the heat capacities C_m and C_H as well as the adiabatic susceptibility χ_S!
4. Discuss, with the above partial results, whether the Curie-behavior of the ideal paramagnet ($T_C = 0$) is compatible with the statement of the third law of thermodynamics!

Exercise 3.9.16 With the results of Exercise 3.9.15 for the ideal paramagnet ($T_C = 0$), discuss the so-called

adiabatic demagnetization.

1. Let the paramagnet be in a heat bath $HB(T_1)$. What an amount of heat ΔQ is removed when the magnetic field is raised from zero to $H \neq 0$?
2. The system is now decoupled from the heat bath and the field is adiabatic-reversibly switched off. Calculate the final temperature!

Exercise 3.9.17 The free energy F of a compressible solid (*model:* elastically coupled *Einstein-oscillators*) has, as function of temperature T and volume V, the following form:

$$F(T, V) = F_0(V) + A T \ln \left(1 - e^{-E(V)/k_B T} \right).$$

For the temperature-independent part we make the ansatz:

$$F_0(V) = \frac{B}{2 V_0} (V - V_0)^2.$$

Furthermore, $E(V)$ is expandable:

$$E(V) = E_0 - E_1 \frac{V - V_0}{V_0}.$$

The quantities A, B, E_0, E_1 are positive constants.

1. Calculate the pressure p, the entropy S and the internal energy U as functions of T and V. Express the results as far as possible by the *Bose function*

$$n(T, V) = \left(e^{E(V)/k_B T} - 1\right)^{-1} .$$

2. Which volume does the body have at vanishing pressure? How large is the thermal expansion coefficient β? Discuss in particular the limiting cases $T = 0$ and $k_B T \gg E(V)$. Thereby restrict yourself to contributions of lowest non-vanishing order in E_1.
3. Estimate to the same accuracy as in 2., the difference $C_p - C_V$ of the heat capacities.

Exercise 3.9.18 The work, which is necessary to enlarge the surface A of a liquid at constant volume by dA, may be given by $\sigma\, dA$ with

$$\sigma = \sigma(T) = \alpha \left(1 - \frac{T}{T_c}\right) \quad (T < T_c; \ \alpha > 0) .$$

$C_{V,A}$ is the heat capacity for simultaneously constant volume and surface.

1. How does the differential dU of the internal energy $U = U(S, V, A)$ read?
2. Prove the relation

$$\left(\frac{\partial T}{\partial A}\right)_{S,V} = \frac{T}{C_{V,A}} \frac{d\sigma}{dT} .$$

3. For an adiabatic-isochoric reversible process calculate the temperature as a function of the surface, with given initial values $T = T_0$, $A = A_0$ and with $C_{V,A}$ as a constant.
4. How does the differential dF of the free energy $F = F(T, V, A)$ read?
5. Show that F decomposes into a *volume part* $F_V(T, V)$ and a *surface part* $F_A(T, A)$.
6. For an isothermal-isochoric process, how large is the change dS of the entropy when the surface changes by dA?
7. How does U change with the surface for an isothermal-isochoric process?
8. What is the surface part $S_A(T, A)$ of the entropy? How much heat is necessary in order to change the surface by an isothermal-isochoric process from A_1 to A_2?
9. Find the differential of the free enthalpy!
10. Calculate the surface part of the free enthalpy! How does one get the volume of the system from G_V?

Exercise 3.9.19

1. A liquid drop (radius r, mass M_1, density ρ_1) is present within the vapor (mass M_2) of the same substance. As in Exercise 3.9.18, one decomposes the free enthalpy into a volume part and a surface part. The volume part per unit mass is

g_1 for the liquid, the free enthalpy per unit mass of the vapor is g_2. Temperature and pressure in both phases are the same. How does the free enthalpy read for the total system? (If necessary, use partial results from Exercise 3.9.18.)

2. For given pressure p and given temperature T the total free enthalpy is minimal at equilibrium. Use this principle to derive the relation

$$g_2 - g_1 = \frac{2\sigma(T)}{r\rho_1}$$

($\sigma(T)$ as in Exercise 3.9.18, $\rho_1 = $ const).

3. Let the density of the vapor ρ_2 be very much smaller than ρ_1. The vapor is assumed to behave like an ideal gas. Under these presumptions derive the vapor pressure of the drop

$$p = p(r, T) \ .$$

Exercise 3.9.20 Consider a system of magnetic moments with the thermodynamic variables: Temperature T, magnetic field H, and magnetization M (pressure p and volume V are constant and irrelevant for the following).

1. The internal energy $U = U(T, M)$ is known and, furthermore, the equation of state is given in the form $M = f(T, H)$. Formulate with these specifications, the difference of the heat capacities $C_M - C_H$.

2. What follows especially for the ideal paramagnet

$$\left[\left(\frac{\partial U}{\partial M}\right)_T = 0 ; \quad M = \frac{C}{T} H ; \quad C : \text{Curie constant} \right] ?$$

3. Prove the following relations:

a) $\left(\frac{\partial S}{\partial M}\right)_T = -\mu_0 V \left(\frac{\partial H}{\partial T}\right)_M$,

b) $\left(\frac{\partial S}{\partial H}\right)_T = \mu_0 V \left(\frac{\partial M}{\partial T}\right)_H$,

c) $\left(\frac{\partial S}{\partial M}\right)_T = \frac{1}{T}\left[\left(\frac{\partial U}{\partial M}\right)_T - \mu_0 V H\right] \ .$

4. Verify with 1. and 3. the assertion:

$$C_M - C_H = \mu_0 V T \left(\frac{\partial H}{\partial T}\right)_M \left(\frac{\partial M}{\partial T}\right)_H \ .$$

5. Use for the calculation of $C_M - C_H$ the following equation of state:

$$H = \frac{1}{C}(T - T_c)M + bM^3 .$$

C, T_c, b are positive constants.

6. Show that in the case of such an equation of state the heat capacity C_M can **not** depend on M.

7. With the equation of state from 5., calculate $F = F(T, M)$ and $S = S(T, M)$.

8. Show that the equation of state in 5. possesses, in a certain temperature region, for $H \rightarrow 0$, besides the self-evident solution $M = 0$, also a non-trivial solution $M = M_S \neq 0$. Discuss the relative stability of the two mathematical solutions by comparing the respective free energies.

9. How do the magnetic susceptibility χ_T and the difference $C_M - C_H$ depend on temperature in the limit $H \rightarrow 0$?

Exercise 3.9.21   One consequence of the third law of thermodynamics is the vanishing of the heat capacities at the absolute zero (3.84), (3.85), for instance

$$\lim_{T \to 0} C_p = 0$$

That suggests the experimentally also confirmed ansatz:

$$C_p = T^x (a + bT + cT^2 + \ldots) \quad x > 0 \, ; \, a = a(p) \neq 0 \, ; \, b = b(p) \, ; \, c = c(p) .$$

Let

$$\beta = \frac{1}{V}\left(\frac{\partial V}{\partial T}\right)_p$$

be the isobaric expansion coefficient.

1. Show that the ratio

$$\frac{V\beta}{C_p}$$

goes for $T \rightarrow 0$ to a finite constant!

2. Prove:

$$\lim_{T \to 0}\left(\frac{\partial T}{\partial p}\right)_S = 0 .$$

What follows herefrom for the attainability of the absolute zero of the temperature scale?

3.10   Self-Examination Questions

To Sect. 3.1

1. Which partial derivatives of $U = U(S, V, \mathbf{N})$ determine the *response functions* C_V and κ_S of a gas?
2. What does one understand by a *Maxwell relation*?
3. Under which conditions is a state function called a *thermodynamic potential*?
4. Does the internal energy U in the form of the caloric equation of state $U = U(T, V, \mathbf{N})$ represent a thermodynamc potential?
5. When do we speak of *natural variables* of a thermodynamic potential?
6. Which are the natural variables of the internal energy U of a gas?

To Sect. 3.2

1. Why is it reasonable to introduce, besides the internal energy U, further thermodynamic potentials?
2. By what do free and internal energy differ from each other?
3. How does the total differential dF of the free energy of a gas with fixed particle number read?
4. By which partial derivative of F is the entropy S given?
5. Formulate the Legendre transformation from the free energy F to the enthalpy H!
6. Which are the natural variables of the free (Gibb's) enthalpy G?
7. What is dG for a magnetic system?

To Sect. 3.3

1. What is the Gibbs-Duhem relation?
2. From which general property of thermodynamic potentials does the Gibbs-Duhem relation result?
3. Which physical meaning can be ascribed to the chemical potential μ if one looks at the Gibbs-Duhem relation?
4. What is meant by *homogeneity relations* of the thermodynamic potentials?

To Sect. 3.4

1. The internal energy $U = U(S, V, N)$ of the ideal gas is volume-dependent. Is this a contradiction to the Gay-Lussac experiment?
2. Sketch the procedure for the calculation of the chemical potential $\mu(T, V, N)$ of the ideal gas!

To Sect. 3.5

1. Describe a reversible auxiliary process for the irreversible intermixing of two gases consisting of non-identical particle types!
2. How does the entropy of mixing ΔS read, which appears with the intermixing of α gases with pairwise different particle types?
3. Describe the problem that arises in connection with the intermixing of gases which consist of particles of the same type?
4. What is called *Gibb's paradox*?

To Sect. 3.6

1. Describe the Joule-Thomson effect!
2. Which thermodynamic potential remains constant during the Joule-Thomson process?
3. How is the *differential Joule-Thomson coefficient* defined? What is its physical meaning?
4. Why can the throttled adiabatic expansion of an ideal gas not lead to a cooling effect?
5. What do we understand by an *inversion curve*?
6. How does the entropy behave during the Joule-Thomson process?

To Sect. 3.7

1. Which thermodynamic potential describes most reasonably the transition of an isolated system into equilibrium?
2. What are the equilibrium conditions for an isolated system? What can be said about temperature, pressure, and chemical potential at equilibrium?
3. Which are the equilibrium conditions for a closed system in a heat bath without any exchange of work? Which thermodynamic potential is relevant?

4. How does the free enthalpy behave in a closed system in a heat bath with constant forces? Which are the equilibrium conditions?
5. To which situations are the extremal properties of U and H suitable?

To Sect. 3.8

1. What is the statement of the third law of thermodynamics?
2. Is the $T \to 0$-limiting value of the entropy dependent on the values of the other variables?
3. What follows for the heat capacities from the third law of thermodynamics?
4. In what way does the behavior of the *ideal gas* violate the third law of thermodynamics?
5. What can be said about the isobaric expansion coefficient β in the limit $T \to 0$?
6. Give reasons for the unattainability of the absolute zero of temperature!

Chapter 4
Phases, Phase Transitions

4.1 Phases

4.1.1 Gibbs Phase Rule

In Sect. 3.7 we have derived the equilibrium conditions for thermodynamic systems. Those considerations can be further generalized. In the form of a *gedanken-experiment*, we divided the total system into two *fictitious* partial systems and therewith created a simple non-equilibrium situation. The system reacted to this situation in a definite manner and provided therewith information about the behavior of certain state quantities at equilibrium. We now realize such a subdivision of the system, without using walls, by different

phases

which coexist in one and the same thermodynamic system. One denotes as *phases* the possible, different forms of state of a macroscopic substance, for instance the different states of aggregation: solid, liquid, gaseous. In the different phases, certain macroscopic observables, as e.g. the particle density, can possess completely different values.—For the following discussion we perform a case-by-case analysis:

A) **Isolated system** Let us assume that this system consists of π phases ($\nu = 1, 2, \ldots, \pi$), where each is composed by α components ($j = 1, 2, \ldots, \alpha$), i.e. there are α different particle types. We write:

$$\sum_{\nu=1}^{\pi} V_\nu = V = \text{const} ,$$

$$\sum_{\nu=1}^{\pi} U_\nu = U = \text{const} ,$$

© Springer International Publishing AG 2017

W. Nolting, *Theoretical Physics 5*, DOI 10.1007/978-3-319-47910-1_4

$$\sum_{\nu=1}^{\pi} N_{j\nu} = N_j = \text{const}_j \; ; \quad j = 1, 2, \dots, \alpha \, .$$

The entropy is an extensive state quantity, therefore

$$S(U, V, \mathbf{N}) = \sum_{\nu=1}^{\pi} S_\nu (U_\nu, V_\nu, \mathbf{N}_\nu) \, . \tag{4.1}$$

We look for the equilibrium state for which, according to (3.70), we must have $dS = 0$, i.e., consistent with of the above boundary conditions the entropy has to become extremal. For the derivation, we utilize the

method of Lagrange multipliers ,

which we have introduced in section 1.2.5 of Vol. 2. We have to require:

$$dS = \sum_{\nu=1}^{\pi} \left\{ \left(\frac{\partial S_\nu}{\partial U_\nu} \right)_{V_\nu, \mathbf{N}_\nu} dU_\nu + \left(\frac{\partial S_\nu}{\partial V_\nu} \right)_{U_\nu, \mathbf{N}_\nu} dV_\nu \right.$$
$$\left. + \sum_{j=1}^{\alpha} \left(\frac{\partial S_\nu}{\partial N_{j\nu}} \right)_{V_\nu, U_\nu, N_{i\nu, i \neq j}} dN_{j\nu} \right\} \stackrel{!}{=} 0 \, . \tag{4.2}$$

Since not all the quantities U_ν, V_ν, $N_{j\nu}$ are independent of each other, we can not simply argue that all coefficients of the dU_ν, dV_ν, $dN_{j\nu}$ have to already vanish. Because of the boundary conditions we can, however, exploit:

$$\sum_\nu dU_\nu = 0 \quad \Longrightarrow \quad \lambda_U \sum_\nu dU_\nu = 0 \, ,$$

$$\sum_\nu dV_\nu = 0 \quad \Longrightarrow \quad \lambda_V \sum_\nu dV_\nu = 0 \, ,$$

$$\sum_\nu dN_{j\nu} = 0 \quad \Longrightarrow \quad \lambda_j \sum_\nu dN_{j\nu} = 0 \, .$$

At first λ_U, λ_V, λ_j are real numbers being not further specified, which are denominated as

Lagrange parameters (multipliers) .

We can now combine the extremal condition for S with the boundary conditions in the following manner:

$$
0 \overset{!}{=} \sum_{v=1}^{\pi} \left[\left(\frac{\partial S_v}{\partial U_v} \right)_{V_v, N_v} - \lambda_U \right] dU_v + \sum_{v=1}^{\pi} \left[\left(\frac{\partial S_v}{\partial V_v} \right)_{U_v, N_v} - \lambda_V \right] dV_v
$$

$$
+ \sum_{v=1}^{\pi} \sum_{j=1}^{\alpha} \left[\left(\frac{\partial S_v}{\partial N_{jv}} \right)_{U_v, V_v, N_{iv, i \neq j}} - \lambda_j \right] dN_{jv} . \tag{4.3}
$$

λ_U, λ_V and λ_j are still freely selectable. Because of the boundary conditions, however, the U_v, V_v, N_{jv} are not independent of each other. For each of the energies U_v, the volumes V_v, and the particle numbers N_{jv}, there exists exactly **one** constraint. We therefore can decompose them into one dependent and $(\pi - 1)$ independent variables, e.g.

$$
\begin{aligned}
&U_1 \text{ dependent;} \quad U_2, \ldots, U_\pi \text{ independent,} \\
&V_1 \text{ dependent;} \quad V_2, \ldots, V_\pi \text{ independent,} \\
&N_{j1} \text{ dependent;} \quad N_{j2}, \ldots, N_{j\pi} \text{ independent.}
\end{aligned}
$$

We now fix the Lagrange parameters λ_U, λ_V, λ_j such that

$$
\left(\frac{\partial S_1}{\partial U_1} \right)_{\ldots} = \lambda_U ; \quad \left(\frac{\partial S_1}{\partial V_1} \right)_{\ldots} = \lambda_V ; \quad \left(\frac{\partial S_1}{\partial N_{j1}} \right)_{\ldots} = \lambda_j .
$$

We thereby achieve that the $v = 1$-summands in (4.3) disappear. The remaining summands, however, then contain only independent variables so that already each square bracket by itself has to become zero. By the use of the multipliers, we have thus come up to the point that it holds now **for all** v:

$$
\left(\frac{\partial S_v}{\partial U_v} \right)_{\ldots} = \frac{1}{T_v} \overset{!}{=} \lambda_U \quad \Longrightarrow \quad T_v = T \quad \forall v , \tag{4.4}
$$

$$
\left(\frac{\partial S_v}{\partial V_v} \right)_{\ldots} = \frac{p_v}{T_v} \overset{!}{=} \lambda_V \quad \Longrightarrow \quad p_v = p \quad \forall v , \tag{4.5}
$$

$$
\left(\frac{\partial S_v}{\partial N_{jv}} \right)_{\ldots} = -\frac{\mu_{jv}}{T_v} \overset{!}{=} \lambda_j \quad \Longrightarrow \quad \mu_{jv} = \mu_j \quad \forall v . \tag{4.6}
$$

The parameters λ_U, λ_V, λ_j are all independent of ν and therewith the same for all phases. This leads to the important conclusion:

In an isolated system at equilibrium all the phases have

1. *the same temperature T,*
2. *the same pressure p,*
3. *the same chemical potential μ_j.*

Let us now apply the same procedure to another experimentally important situation:

B) Closed System with $p = $ const, $T = $ const Pressure and temperature are fixed from the outside. That is the situation discussed in Sect. 3.7.3. Because of (3.76):

$$\text{equilibrium} \Longleftrightarrow dG = 0 ; \quad G \text{ minimal!}$$

The free enthalpy is an extensive state quantity:

$$G = \sum_{\nu=1}^{\pi} G_\nu (T, p, \mathbf{N}_\nu) . \tag{4.7}$$

We assume a free interchangeability of particles between the various phases while, however, the total number of particles does not change:

$$N_j = \sum_{\nu=1}^{\pi} N_{j\nu} \quad \Longrightarrow \quad \lambda_j \sum_{\nu=1}^{\pi} dN_{j\nu} = 0 . \tag{4.8}$$

We couple this boundary condition to the extremal condition for G by the use of the Lagrange parameter λ_j:

$$0 \overset{!}{=} \sum_{\nu=1}^{\pi} \sum_{j=1}^{\alpha} \left[\left(\frac{\partial G_\nu}{\partial N_{j\nu}} \right)_{N_{i\nu}, i \neq j}^{T, p} - \lambda_j \right] dN_{j\nu} . \tag{4.9}$$

A similar conclusion as that in A) now leads to:

$$\left(\frac{\partial G_\nu}{\partial N_{j\nu}} \right)_{N_{i\nu}, i \neq j}^{T, p} = \mu_{j\nu} = \lambda_j . \tag{4.10}$$

Since λ_j is independent of ν, we come to the important result:

In a closed system with $p = $ const and $T = $ const, which is at equilibrium, the chemical potential of the particle type j has in all the phases the same value:

$$\mu_{j\nu} \equiv \mu_j \quad \forall \nu . \tag{4.11}$$

Let us evaluate this result a bit further. Formally it holds actually:

$$\left(\frac{\partial G_\nu}{\partial N_{j\nu}} \right)_{N_{i\nu}, i \neq j}^{T, p} = \mu_{j\nu} = \mu_{j\nu} (T, p; N_{11}, \dots, N_{\alpha\pi}) .$$

Since the μ_{jv} are intensive quantities there cannot be a direct dependence on the extensive variables N_{jv}. The chemical potentials μ_{jv} will in reality depend on the **concentrations** c_{jv},

$$c_{jv} = \frac{N_{jv}}{N_v} \; ; \quad \sum_{j=1}^{\alpha} c_{jv} = 1 \; , \tag{4.12}$$

which are of course intensive quantities:

$$\mu_{jv} = \mu_{jv}\,(T, p; c_{11}, \ldots, c_{\alpha\pi}) \; . \tag{4.13}$$

In the argument we find Z_V variables,

$$Z_V = 2 + \alpha\,\pi \; , \tag{4.14}$$

which, however, are not independent of each other since a series of boundary conditions are to be fulfilled. The relation (4.12) yields, because of $v = 1, 2, \ldots, \pi$,

$$Z_N^{(1)} = \pi$$

constraints. The equilibrium condition (4.11) leads for each j to $(\pi - 1)$ equations between the μ_{jv}. This gives

$$Z_N^{(2)} = \alpha\,(\pi - 1)$$

further constraints. Let us define

$$f = \text{number of the degrees of freedom,}$$
$$= \text{number of the independently selectable variables,}$$

for which we have obviously:

$$f = Z_V - Z_N^{(1)} - Z_N^{(2)} = 2 + \alpha\,\pi - \pi - \alpha\,(\pi - 1) \; .$$

This yields the important

Gibbs Phase Rule

$$f = 2 + \alpha - \pi \; , \tag{4.15}$$

where

$$\alpha = \text{number of components,}$$
$$\pi = \text{number of phases} \; .$$

Fig. 4.1 Phase diagram of water

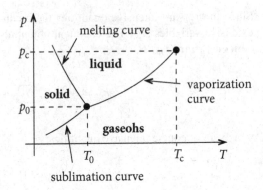

sublimation curve

We illustrate this phase rule by a well-known example:

H₂O-Phase Diagram

$$T_0 = 0.0075\,^{\circ}\text{C} \ \textbf{triple point},$$
$$T_c = 374.2\,^{\circ}\text{C} \ \ \textbf{critical point}.$$

It is about a one-component system, i.e., $\alpha = 1$ (Fig. 4.1).

1) $\pi = 1$

In the one-phase regions (solid, liquid, gaseous) we find

$$f = 2\,,$$

i.e., p and T can still be chosen independently.

2) $\pi = 2$

On the curves of coexistence we have

$$f = 1\,,$$

so that only one variable can freely be chosen, e.g. T, the other is then fixed, e.g. $p = p(T)$.

3) $\pi = 3$

At the triple point (T_0, p_0) the three phases are all at equilibrium with one another. There is not any freely selectable parameter left:

$$f = 0\,.$$

Out of the phase rule (4.15) we can draw a further conclusion, namely, that obviously there exists an upper limit for the number π of the possible phases,

$$\pi \leq 2 + \alpha\,, \tag{4.16}$$

since f can not of course become negative.

4.1.2 Vapor Pressure Curve (Clausius-Clapeyron)

Let us discuss, as an example of application of the equilibrium conditions of the preceding section, a simple but important special case (see also Exercise 2.9.29), which concerns the equilibrium between liquid (l) and vapor (g) of a one-component system like, e.g., H_2O.

If one chooses p and T as the variables, then according to (4.13) at equilibrium we have:

$$\mu_l(T,p) = \mu_g(T,p) . \tag{4.17}$$

From this relation, it must in principle be possible to derive a relation $p = p(T)$ for those states in which liquid and vapor (gas) are at equilibrium with each other.

If, however, (4.17) is not valid, then the Gibbs-Duhem-Relation (3.34),

$$G(T,p,N) = N\mu(T,p) ,$$

tells us that the equilibrium is completely shifted to the phase with the smaller μ. The phase with the minimal free enthalpy is stable (Fig. 4.2)!

According to (3.26) we have:

$$dG - \mu\, dN = -S\, dT + V\, dp = N\, d\mu(T,p) .$$

We consider a *shift* (dp, dT) along the line of coexistence (**vapor pressure curve**). There we have because of (4.17)

$$d\mu_l(T,p) = d\mu_g(T,p)$$

and therewith ($N_l = N_g = N$):

$$-S_l\, dT + V_l\, dp = -S_g\, dT + V_g\, dp .$$

We get herefrom the slope dp/dT of the vapor pressure curve:

$$\frac{dp}{dT} = \frac{S_g - S_l}{V_g - V_l} . \tag{4.18}$$

Fig. 4.2 Chemical potential as function of the temperature for the gaseous and the liquid phases of water (schematic!)

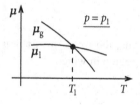

Usually this expression is referred to one mole:

$v_{g,l}$:   mole volume for, respectively, gas and liquid,

$s_{g,l}$:   entropies per mole.

One eventually defines:

$$Q_M = T(s_g - s_l) : \quad \textbf{molar heat of evaporation.}$$

This part is needed to overcome the cohesive forces between the particles. By (4.18) we finally get the
Clausius-Clapeyron equation

$$\frac{dp}{dT} = \frac{Q_M}{T\left(v_g - v_l\right)} .\tag{4.19}$$

For the derivation of (4.19) and (4.18), we had to implicitly presume that $S_g \neq S_l$ and $V_g \neq V_l$. This means:

$$\mu_l(T,p) \overset{!}{=} \mu_g(T,p) ,$$

$$\left(\frac{\partial \mu_l}{\partial T}\right)_p \neq \left(\frac{\partial \mu_g}{\partial T}\right)_p \quad ; \quad \left(\frac{\partial \mu_l}{\partial p}\right)_T \neq \left(\frac{\partial \mu_g}{\partial p}\right)_T .\tag{4.20}$$

Such a transition '*gas* $\Longleftrightarrow$ *liquid*' is called a

phase transition of first order

The Clausius-Clapeyron equation is valid only for such transitions!

4.1.3 Maxwell Construction

We observed already in Sect. 1.4.2, when we discussed the equation of state of the van der Waals gas,

$$\left(p + a\frac{n^2}{V^2}\right)(V - nb) = nRT ,$$

that the isotherms exhibit an unphysical behavior for $T < T_c$. There is namely a region in which

$$\kappa_T = -\frac{1}{V}\left(\frac{\partial V}{\partial p}\right)_T < 0 .$$

Fig. 4.3 Isotherms of the real gas in the pV-diagram with the 'Maxwell construction' for $T < T_c$

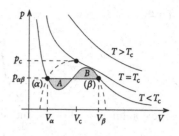

That means the system would be mechanically unstable. The reason lies in the implicit assumption that, the van der Waals 'gas' for all parameter constellations, consists of only one single homogeneous phase. In reality, however, in the region mentioned, there are two (!) phases at equilibrium with each other, namely, liquid and vapor (gas). The van der Waals isotherms loose their validity in the two-phase region.

The considerations after (4.17) have shown that in the region of coexistence of gas and liquid, the pressure can be a function of the temperature only, in particular being independent of the volume. This means:

*All isotherms $(T < T_c)$ run **horizontally** in the pV-diagram!*

At the one end (α) (see Fig. 4.3) of this horizontal piece of the isotherm only liquid is present, at the other end (β) only gas. To the left of (α) and to the right of (β) we can again use the van der Waals equation.

In the region of coexistence, the equilibrium condition

$$\mu_l\left(T,p_{\alpha\beta}\right) = \mu_g\left(T,p_{\alpha\beta}\right) = \text{const} \tag{4.21}$$

must be fulfilled. At the edges (α) and (β) **all** the N particles are in the liquid and in the gaseous phase, respectively. With the Gibbs-Duhem relation (3.34) it follows then from (4.21):

$$G_\alpha\left(T,p_{\alpha\beta}\right) = G_\beta\left(T,p_{\alpha\beta}\right) ,$$
$$U_\alpha - T\,S_\alpha + p_{\alpha\beta}\,V_\alpha = U_\beta - T\,S_\beta + p_{\alpha\beta}\,V_\beta . \tag{4.22}$$

For the difference in the **free energies** it results herefrom:

$$F_\alpha - F_\beta = p_{\alpha\beta}\left(V_\beta - V_\alpha\right) . \tag{4.23}$$

But if one takes with the original van der Waals pressure $p = p(T,V)$ the integral from V_α to V_β, it follows, because of

$$dF = -p\,dV \quad \text{at} \quad T = \text{const} ,$$

for the free energy:

$$F_\alpha - F_\beta = -\int_{V_\beta}^{V_\alpha} p\,dV = \int_{V_\alpha}^{V_\beta} p\,dV \quad (T = \text{const}) . \tag{4.24}$$

The combination of (4.23) and (4.24)

$$\int_{V_\alpha}^{V_\beta} p\,dV = p_{\alpha\beta}(T)\left(V_\beta - V_\alpha\right) \quad (T = \text{const}) \tag{4.25}$$

permits then a simple geometrical interpretation. In Fig. 4.3 the indicated areas A and B must be equal:

$$A \overset{!}{=} B : \qquad \textbf{Maxwell construction!}$$

Therewith we have been able to derive, from the general equilibrium conditions, a prescription for how to transform the van der Waals isotherms into **physical** isotherms everywhere.

At the end, we can further convince ourselves easily that the two-phase region, which dissociates into gas and liquid, is energetically more stable than the original van der Waals-one-phase region. For this purpose we compare two states with the same T and V, but different pressures, $p_{\alpha\beta}$ for the two-phase system and $p(T, V)$ corresponding to the van der Waals equation (Fig. 4.4). Since T and V are preset, the free energy F must be minimal at the equilibrium!

Fig. 4.4 Interpretation of the two-phase region and the respective Maxwell construction via the minimum condition concerning the free energy F

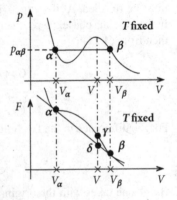

In the **one-phase region** we have:

$$F_{vdW}(T, V) - F_\alpha = F_\gamma - F_\alpha$$

$$= -\int_{V_\alpha}^{V} p(T, V')\, dV' . \tag{4.26}$$

According to the Maxwell construction the free energy in the **two-phase region** is composed by the parts of the liquid (α) and the gas (β), respectively:

$$F(T, V) = F_\delta = c_1 F_\alpha(T, V_\alpha) + c_g F_\beta(T, V_\beta) .$$

Thereby it must be:

$$c_1 = \frac{N_1}{N} ; \quad c_g = \frac{N_g}{N} \quad \Longrightarrow \quad c_1 + c_g = 1 .$$

N_1, N_g are the particle numbers of, respectively, the liquid and the gaseous phase. Both phases together must of course fill the volume V:

$$V = c_1 V_\alpha + c_g V_\beta .$$

The last two expressions lead to the so-called **lever rule** (Fig. 4.4):

$$c_1 = \frac{V_\beta - V}{V_\beta - V_\alpha} ; \quad c_g = \frac{V - V_\alpha}{V_\beta - V_\alpha} . \tag{4.27}$$

This means for the free energy in the two-phase region:

$$F_\delta - F_\alpha = c_g \left(F_\beta - F_\alpha\right) \overset{(4.23)}{=} c_g \left(-p_{\alpha\beta}\right) \left(V_\beta - V_\alpha\right) ,$$
$$F_\delta - F_\alpha = p_{\alpha\beta} \left(V_\alpha - V\right) . \tag{4.28}$$

Combining (4.28) and (4.26) one arrives at:

$$F_\gamma - F_\delta = p_{\alpha\beta} \left(V - V_\alpha\right) - \int_{V_\alpha}^{V} p(T, V')\, dV' \geq 0 . \tag{4.29}$$

Hence, the two-phase region is stable:

$$F_\delta \leq F_\gamma . \tag{4.30}$$

The equal sign is valid only for $V = V_\alpha$ and $V = V_\beta$, respectively.

4.2 Phase Transitions

4.2.1 Geometrical Interpretation

We consider once more the liquid-gas mixture along the line of coexistence. We had seen that the transition takes place at constant temperature and constant pressure:

$$p_{\alpha\beta}(T): \quad \textbf{vapor pressure}$$

In the transition region we have a mixture of liquid in the state (α) and gas in the state (β) (Fig. 4.5). The relative portions are to be determined by the *lever rule* (4.27). In the two-phase region a heat supply leads to conversion of a certain amount of liquid into vapor. The process runs isothermally, since the heat energy is spent exclusively to overcome the particle bindings. Only after reaching the state (β), where all the liquid has been converted into vapor, any further heat supply is used to increase the temperature. Such a phase transition, which requires a

latent heat (transformation heat),

is called a

phase transition of the first order.

For such transitions, the Clausius-Clapeyron equation in the form of (4.18) holds, which is reasonable obviously only if the entropies and the volumes of the vapor (gas) (S_g, V_g) and liquid (S_l, V_l) are different on the line of coexistence. We remember that S and V are **first** partial derivatives of the free enthalpy $G(T, p)$ with respect to T and p, respectively:

$$S = -\left(\frac{\partial G}{\partial T}\right)_p ; \quad V = \left(\frac{\partial G}{\partial p}\right)_T .$$

It is therefore typical for phase transitions of the first order that the first derivatives of $G(T, p)$ are discontinuous for the transition across the line of coexistence.

We try to illustrate the subject a bit in a geometrical manner. For that, however, we have to make some preparations.

Fig. 4.5 Actual isotherm of the real gas

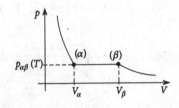

Fig. 4.6 Example of a convex function

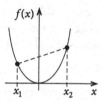

Definition 4.2.1   One calls $f(x)$ a

convex function of x

if one finds for arbitrary λ with $0 \le \lambda \le 1$:

$$f(\lambda x_1 + (1 - \lambda)x_2) \le \lambda f(x_1) + (1 - \lambda)f(x_2) \ . \qquad (4.31)$$

As an example, in Fig. 4.6, the **convex** function $f(x) = x^2$ is plotted.—The chord, which interconnects the points $f(x_1)$ and $f(x_2)$ of a convex function $f(x)$, always lies above or on the curve $f(x)$ for all x in the interval $x_1 \le x \le x_2$. Each tangent to $f(x)$ then lies completely below $f(x)$.—The definition does **not** imply the differentiability of the respective function. If, however, $f(x)$ is two-times differentiable then it further holds:

$$f(x) \quad convex \quad \Longleftrightarrow \quad f''(x) \ge 0 \quad \forall x \ . \qquad (4.32)$$

Completely analogously one defines:

Definition 4.2.2 $\tilde{f}(x)$ is a

concave function

if $(-\tilde{f}(x))$ is convex.

By means of these definitions we can now present statements on the **geometrical** behavior of the free enthalpy and the free energy:

Assertion 4.2.1

1. The free enthalpy $G(T, p)$ is concave with respect to both the variables T and p!
2. The free energy $F(T, V)$ is, as function of T concave and as function of V convex!

The **proof** exploits the so-called **stability conditions**:

$$\text{thermal:} \quad C_x \ge 0 \ ; \quad x = V, p \ , \qquad (4.33)$$

$$\text{mechanical:} \quad \kappa_y \ge 0 \ ; \quad y = S, T \ , \qquad (4.34)$$

which can be proven, in a strict sense, only in the framework of Statistical Mechanics (Vol. 8), although they are illustratively completely evident. $C_x < 0$ would mean that a heat supply causes a lowering of temperature. $\kappa_y < 0$ would mean $(\partial V/\partial p)_y > 0$. With decreasing volume the pressure of a thermodynamic system would also become smaller. Such a system would therefore be mechanically unstable and would implode.

Proof

1): Except for points of phase transitions $G(T,p)$ is differentiable:

$$\left(\frac{\partial^2 G}{\partial T^2}\right)_p = -\left(\frac{\partial S}{\partial T}\right)_p = -\frac{C_p}{T} \leq 0 \,.$$

So $G(T,p)$ is concave as function of T!

$$\left(\frac{\partial^2 G}{\partial p^2}\right)_T = \left(\frac{\partial V}{\partial p}\right)_T = -V\kappa_T \leq 0 \,.$$

So $G(T,p)$ is also as function of p concave (Fig. 4.7)!

2):

$$\left(\frac{\partial^2 F}{\partial T^2}\right)_V = -\left(\frac{\partial S}{\partial T}\right)_V = -\frac{C_V}{T} \leq 0 \,.$$

$F(T,V)$ is indeed as function of T concave!

$$\left(\frac{\partial^2 F}{\partial V^2}\right)_T = -\left(\frac{\partial p}{\partial V}\right)_T = +\frac{1}{V\kappa_T} \geq 0 \,.$$

$F(T,V)$ is as function of V convex (Fig. 4.7)!

When transferring the above assertion to magnetic systems one has to pay attention a bit because the susceptibility χ as analog of the compressibility κ, in contrast to κ, can become even negative (diamagnetism!).

Using these general properties of G and F as well as the relationship

$$G = F + pV$$

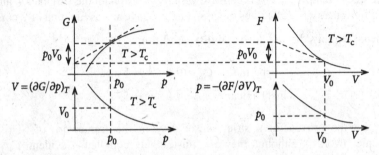

Fig. 4.7 Geometrical construction of the free energy and its volume-derivative out of the free enthalpy and its pressure-derivative in the case that there are no phase transitions

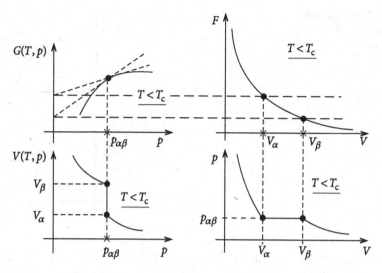

Fig. 4.8 The same as in Fig. 4.7 but now with a phase transition of first order

the dependencies of the potentials on T and p or T and V can already qualitatively be sketched. The points of phase transition are of course of special interest (Fig. 4.8):

The coexistence of the two phases at $(T < T_c, p = p_{\alpha\beta})$ has the consequence that $G_l(T, p_{\alpha\beta}) = G_g(T, p_{\alpha\beta})$. That means that the free energy F must be a linear function of V between V_α and V_β. The phase transition of first order manifests itself in a discontinuity for the first derivative of G with respect to p, i.e. for the volume V, and in a horizontal segment of the first derivative of F with respect to V, i.e. for the pressure p (Fig. 4.8).

As functions of T the potentials F and G behave qualitatively very similarly. At a phase transition of the first order, both functions exhibit, because of

$$S(T, p) = -\left(\frac{\partial G}{\partial T}\right)_p \qquad S(T, V) = -\left(\frac{\partial F}{\partial T}\right)_V , \qquad (4.35)$$

a finite jump in the first partial derivative, i.e. in the entropy (Fig. 4.9):

$$\Delta S = S_\beta - S_\alpha .$$

A typical feature of the phase transition of first order is therefore the appearance of a
latent heat (transformation heat, evaporation heat)

$$\Delta Q = T_{\alpha\beta} \, \Delta S \qquad (4.36)$$

Fig. 4.9 Behavior of the
entropy at a phase transition
of first order

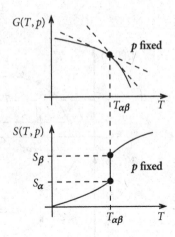

However, ΔQ is not a material constant. In fact, one observes, e.g. for the gas-liquid system, that with a modification of the system parameters, e.g. the fixed pressure p, the discontinuity jump in the first derivatives of the thermodynamical potential $G(T,p)$ changes. If one approaches on the line of coexistence (*vaporization curve*, see H_2O-phase diagram in Fig. 4.1) the critical point, the discontinuity becomes smaller and smaller and eventually completely vanishes at (T_c, p_c). Thus there are also phase transitions with $S_\alpha = S_\beta$ and $V_\alpha = V_\beta$, for which the Clausius-Clapeyron equation (4.19) becomes meaningless. Therefore we obviously have further to extend the term *phase transition* beyond what has been said so far.

4.2.2   Ehrenfest's Classification

Let the phases which are at equilibrium on the line of coexistence again be marked by the indexes α and β. According to Ehrenfest (1933) one defines as the

order of the phase transition

the order of the lowest differential quotient of G which shows a discontinuity at the transition across the line of coexistence. That means explicitly:

Definition 4.2.3 (Phase Transition of n-th Order)

$$1) \quad \left(\frac{\partial^m G_\alpha}{\partial T^m}\right)_p = \left(\frac{\partial^m G_\beta}{\partial T^m}\right)_p \quad \text{for} \quad m = 1, 2, \ldots, n-1, \quad (4.37)$$

$$\left(\frac{\partial^m G_\alpha}{\partial p^m}\right)_T = \left(\frac{\partial^m G_\beta}{\partial p^m}\right)_T \quad \text{for} \quad m = 1, 2, \ldots, n-1, \quad (4.38)$$

$$2) \quad \left(\frac{\partial^n G_\alpha}{\partial T^n}\right)_p \neq \left(\frac{\partial^n G_\beta}{\partial T^n}\right)_p, \quad (4.39)$$

$$\left(\frac{\partial^n G_\alpha}{\partial p^n}\right)_T \neq \left(\frac{\partial^n G_\beta}{\partial p^n}\right)_T . \tag{4.40}$$

Of practical interest are actually only the phase transitions of first and second order. Those of first order have already been analyzed in detail. For a

phase transition of second order

one observes:

1. $G(T,p)$ continuous!
2. $S(T,p); V(T,p)$ continuous!
3. *Response*-functions (Fig. 4.10)

$$C_p = -T\left(\frac{\partial^2 G}{\partial T^2}\right)_p ; \quad \kappa_T = -\frac{1}{V}\left(\frac{\partial^2 G}{\partial p^2}\right)_T ; \quad \beta = \frac{1}{V}\left(\frac{\partial^2 G}{\partial p\, \partial T}\right)$$

discontinuous!

It is immediately clear that with increasing order of the phase transition, the differences of the co-existing phases become physically more and more unimportant. Indeed one has to question up to which order can one reasonably speak of two **different** phases.

Examples for a phase transition of second order in the strict *Ehrenfest sense*, characterized by a **finite** jump of the heat capacity, are not so numerous (Fig. 4.11):

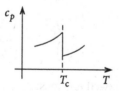

Fig. 4.10 Phase transition of second order according to Ehrenfest's classification, observed by the temperature dependence of the heat capacity

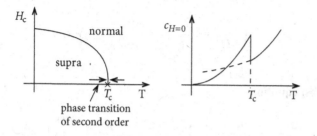

Fig. 4.11 The critical field of a superconductor (*left*) as function of the temperature and the finite zero-field jump of the heat capacity (*right*) at the Ehrenfest-phase transition of second order

1) Models

a) Weiss ferromagnet,
b) Bragg-Williams model (for the order-disorder transition of β-brass),
c) van der Waals-Gas.

2) Superconductor in the zero field

Let us finally derive the analog of the Clausius-Clapeyron equation for second-order phase transitions in the *Ehrenfest scheme*:

On the *line of coexistence*, provided such a line really exists at all, it now holds:

$$S_\alpha(T,p) = S_\beta(T,p) \; ; \quad V_\alpha(T,p) = V_\beta(T,p) \, .$$

Let us consider a change of state (dp, dT) along the *line of coexistence*:

$$dS_\alpha = dS_\beta \; ; \quad dV_\alpha = dV_\beta$$

$$\Longrightarrow \quad \left(\frac{\partial S_\alpha}{\partial T}\right)_p dT + \left(\frac{\partial S_\alpha}{\partial p}\right)_T dp = \left(\frac{\partial S_\beta}{\partial T}\right)_p dT + \left(\frac{\partial S_\beta}{\partial p}\right)_T dp \, ,$$

$$\left(\frac{\partial V_\alpha}{\partial T}\right)_p dT + \left(\frac{\partial V_\alpha}{\partial p}\right)_T dp = \left(\frac{\partial V_\beta}{\partial T}\right)_p dT + \left(\frac{\partial V_\beta}{\partial p}\right)_T dp \, .$$

This can be resolved as follows:

$$\frac{dp}{dT} = -\frac{\left(\frac{\partial S_\beta}{\partial T}\right)_p - \left(\frac{\partial S_\alpha}{\partial T}\right)_p}{\left(\frac{\partial S_\beta}{\partial p}\right)_T - \left(\frac{\partial S_\alpha}{\partial p}\right)_T} = -\frac{\left(\frac{\partial V_\beta}{\partial T}\right)_p - \left(\frac{\partial V_\alpha}{\partial T}\right)_p}{\left(\frac{\partial V_\beta}{\partial p}\right)_T - \left(\frac{\partial V_\alpha}{\partial p}\right)_T} \, .$$

We insert the following Maxwell relations:

$$\left(\frac{\partial S}{\partial p}\right)_T = -\left(\frac{\partial V}{\partial T}\right)_p = -V\beta \, ,$$

$$\left(\frac{\partial S}{\partial T}\right)_p = \frac{C_p}{T} \; ; \quad \left(\frac{\partial V}{\partial p}\right)_T = -V\kappa_T \, .$$

That yields the so-called
Ehrenfest equations

$$\frac{dp}{dT} = \frac{1}{TV} \frac{C_p^{(\alpha)} - C_p^{(\beta)}}{\beta^{(\alpha)} - \beta^{(\beta)}} = \frac{\beta^{(\alpha)} - \beta^{(\beta)}}{\kappa_T^{(\alpha)} - \kappa_T^{(\beta)}} \, . \tag{4.41}$$

It is said that a system shows a phase transition of second order in the *pure Ehrenfest sense*, if $S_\alpha = S_\beta$ and $V_\alpha = V_\beta$ and furthermore if the equations (4.41) are fulfilled.

The today's criticism on the Ehrenfest classification of phase transitions results, on the one hand, from the experimental observation that in many systems with phase transitions, which are definitely **not** of first order, the *critical* thermodynamic quantities exhibit singularities rather than finite jumps. On the other hand, the classification into phase transitions of arbitrarily high order appears to be meaningless!

Today one therefore distinguishes actually, a bit loosely, only two types of phase transitions, which are most effectively demarcated by the behavior of the entropy. S can behave at the transition point either

continuously (2)

or

discontinuously (1)

as a function of the intensive variable T (Fig. 4.12).

1) Discontinuous phase transition This is the already discussed

phase transition of first order

which is characterized by discontinuities in the first partial derivatives of the free enthalpy G (Fig. 4.13).

1. $\Delta S \neq 0 \quad \Longleftrightarrow$ transformation (latent) heat: $\Delta Q = T_0 \, \Delta S$.
2. $C_p = -T \left(\frac{\partial^2 G}{\partial T^2} \right)_p$: finite for $T \neq T_0$, but not defined for $T = T_0$.

The transition temperature T_0 is not at all a constant, but rather dependent on p and V, for fluids, and on $B_0 = \mu_0 H$ and M for magnetic systems, respectively. One observes now for most of the interesting systems a decrease of the discontinuities ΔS, $\Delta \rho$, and $2 M_S$ with increasing T_0 (Fig. 4.14). That defines a

Fig. 4.12 Continuous and discontinuous phase transitions based on the temperature behavior of the entropy

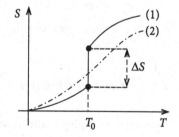

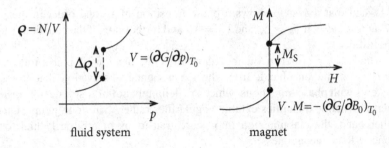

fluid system   magnet

Fig. 4.13 Discontinuous phase transition (first order) for the fluid system (*left*) and the magnet (*right*)

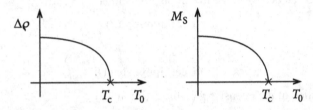

Fig. 4.14 Transition from a phase transition of first order to the one of second order for the fluid system (*left*) and for the magnets (*right*)

critical temperature T_c

at which the first partial derivatives become again continuous. This leads to the other type of *phase transition*!

2) Continuous phase transition In this case one speaks also of a

phase transition of second order

with the following typical characteristics:

1. S continuous $\Longrightarrow \Delta Q = 0$.
2. $T_0 \rightarrow T_c$: *critical point*.
3. singularities in C_V, κ_T, χ_T, i.e. in all state quantities which are second derivatives of the thermodynamic potentials.

In the next sections we will further analyze the continuous phase transition.

4.2.3 Critical Exponents

The **continuous phase transitions** or **phase transition of second order**, defined in the last section, are of special physical interest. The reason for that is an astonishing universality of certain physical properties close to the critical points. Completely different properties of rather completely different systems show, close to T_c, a quasi-unique **power-law behavior**. That will be investigated to some detail in this section. An *in-depth* discussion, though, must be reserved to the Statistical Mechanics in Vol. 8.

In the range of the so-called **critical fluctuations**, which are typically to be expected in the temperature region

$$|\varepsilon| = \left| \frac{T - T_{\mathrm{c}}}{T_{\mathrm{c}}} \right| < 10^{-2} , \tag{4.42}$$

one observes the above-mentioned universal behavior, even for drastically different physical quantities, which can be described by a set of only a few numerical values which one calls the

critical exponents .

Very often one realizes that a

physical property f(ε)

in the critical region (4.42) behaves like

$$f(\varepsilon) = a\,\varepsilon^{\varphi} \left(1 + b\,\varepsilon^{\psi} + \ldots \right) ; \quad \psi > 0 . \tag{4.43}$$

This is expressed in shorthand as

$$f(\varepsilon) \simeq \varepsilon^{\varphi} \tag{4.44}$$

being read as: 'For $\varepsilon \to 0\ f(\varepsilon)$ behaves like ε^{φ}'. φ is then the critical exponent. In the meantime one has, however, found out that the power-law behavior is a bit too restrictive. The definition of the critical exponents is therefore generalized as follows.

Definition 4.2.4 (Critical Exponent)

$$\varphi = \lim_{\varepsilon \to 0} \frac{\ln |f(\varepsilon)|}{\ln |\varepsilon|} . \tag{4.45}$$

The special behavior (4.43) is of course included in this more general definition. It is clear that there is not only one single exponent for all physical quantities, but rather a full set, which we will introduce in detail.

The critical exponents introduced by (4.45) are

almost universal,

i.e., they depend only on the following few criteria:

1. dimension d of the system,
2. range of the particle interactions,
3. spin dimensionality n.

This is the so-called **universality hypothesis** (R. B. Griffiths, Phys. Rev. Lett. **24**, 1479 (1970)), which can be commented on in detail only later in the framework of Statistical Mechanics. The ranges of the particle interactions are grouped into three classes. They are called

short range

if the decay of the interaction strength with the particle distance follows

$$r^{-(d+2+\alpha)} \; ; \quad \alpha > 0 .$$

Details of the particle interactions then do not play any role. One registers a real universal behavior.—The interactions are classified as

long range

if

$$\alpha < \frac{d}{2} - 2 . \tag{4.46}$$

In this case the so-called *classical theories* become valid (Landau theory, van der Waals model, Weiss ferromagnet). These invalidate point 1., i.e. the exponents are not dependent on the dimension d of the system.—The intermediate region appears fairly complicated. For

$$\frac{d}{2} - 2 < \alpha < 0 \tag{4.47}$$

one speaks of an

intermediate range

interaction. The exponents then depend (normally) on α.

Magnetic materials are very often modelled as interacting spin systems. By **spin dimensionality** one understands the number of relevant components of the spin vectors. For $n = 1$ (*Ising model*) we have one-dimensional vectors, for $n = 2$ (*XY model*) two-dimensional vectors and for $n = 3$ (*Heisenberg model*) three-dimensional vectors. The critical exponents turn out to be strongly n-dependent. We will get to know further details about spin systems within the framework of Statistical Mechanics (Vol. 8).

As to the critical exponents, one has to, strictly speaking, distinguish from which side the critical point is approached:

$$\varphi \quad \longleftrightarrow \quad \varepsilon > 0 \quad \left(T \overset{>}{\to} T_{\mathrm{c}} \right) ,$$

$$\varphi' \quad \longleftrightarrow \quad \varepsilon < 0 \quad \left(T \overset{<}{\to} T_{\mathrm{c}} \right) .$$

It is not necessary that $\varphi = \varphi'$. The scaling law hypothesis, which will be discussed later, though, will postulate the equality. We now present some typical examples:

1. $\varphi < 0$

$f(\varepsilon)$ diverges for $\varepsilon \to 0$, and that the divergence is the *sharper* the smaller is $|\varphi|$ ($|\varphi_2| > |\varphi_1|$ in Fig. 4.15). Note that in the interesting region $|\varepsilon| < 1$.

2. $\varphi > 0$

$f(\varepsilon)$ approaches zero for $\varepsilon \to 0$. In the example sketched in Fig. 4.16, $\varphi_1 > \varphi_2$.

3. $\varphi = 0$

In this case the behavior of $f(\varepsilon)$ is not unique. One has to distinguish three different situations:

3.1 Logarithmic divergence

As for instance in Fig. 4.17, we look at

$$f(\varepsilon) = a \ln|\varepsilon| + b ,$$

and get then with (4.45):

$$\varphi = \lim_{\varepsilon \to 0} \frac{\ln|a \ln|\varepsilon| + b|}{\ln|\varepsilon|} = \lim_{\varepsilon \to 0} \frac{\ln|\ln|\varepsilon||}{\ln|\varepsilon|} = \lim_{\varepsilon \to 0} \frac{1}{|\varepsilon|} \frac{\frac{1}{|\ln|\varepsilon||}}{\frac{1}{|\varepsilon|}} = 0 .$$

Fig. 4.15 Critical behavior of a function f with a negative critical exponent as a function of the reduced temperature $(T - T_c) / T_c$

Fig. 4.16 Critical behavior of a function f with a positive critical exponent as a function of the reduced temperature $(T - T_c) / T_c$

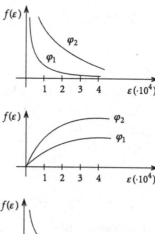

Fig. 4.17 Logarithmical divergence as a possibility for a critical exponent $\varphi = 0$

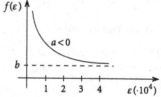

Fig. 4.18 Critical behavior as divergence of the j-th derivative

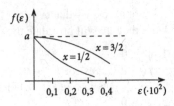

3.2 Divergent j-th derivative

$f(\varepsilon)$ itself can remain finite for $\varepsilon \to 0$, while the j-the derivative diverges, as for instance

$$f(\varepsilon) = a - b\,\varepsilon^x \quad \text{with} \quad x = \frac{3}{2}, \frac{1}{2}.$$

In the examples sketched in Fig. 4.18, for $x = 3/2$ the second and for $x = 1/2$ the first derivative of f diverges when $\varepsilon \to 0$. The critical exponent φ is, however, zero:

$$\varphi = \lim_{\varepsilon \to 0} \frac{\ln|a - b\,\varepsilon^x|}{\ln|\varepsilon|} = \ln|a| \lim_{\varepsilon \to 0} \frac{1}{\ln|\varepsilon|} = 0 \,.$$

In order to distinguish the cases 3.1 and 3.2 one sometimes introduces a new type of critical exponent. If j is the smallest integer for which

$$\frac{\partial^j f}{\partial \varepsilon^j} \equiv f^{(j)}(\varepsilon) \xrightarrow[\varepsilon \to 0]{} \infty \,, \tag{4.48}$$

then it shall be

$$\varphi_S = j + \lim_{\varepsilon \to 0} \frac{\ln|f^{(j)}(\varepsilon)|}{\ln|\varepsilon|} \,. \tag{4.49}$$

For the above examples we have besides $\varphi = 0$:

$$\varphi_S = j - \frac{1}{2} = 1 - \frac{1}{2} = \frac{1}{2} \quad \text{for} \quad x = \frac{1}{2} \,,$$

$$\varphi_S = j - \frac{1}{2} = 2 - \frac{1}{2} = \frac{3}{2} \quad \text{for} \quad x = \frac{3}{2} \,.$$

A logarithmically divergent function has of course $j = 0$ and therewith $\varphi_S = 0$.

3.3 Discontinuities

The function behaves analytically for $T \neq T_c$ with a finite jump at T_c like a phase transition of second order *in the Ehrenfest sense*. In this case, too, it is $\varphi = 0$ (see Exercise 4.3.3)!

Fig. 4.19 Isotherms of the real gas in the pressure-density diagram for defining the paths on which the critical exponents are defined

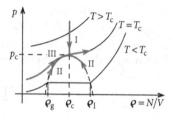

Fig. 4.20 Isotherms of the ferromagnet for fixing the paths on which the critical exponents are defined (T_C: Curie temperature)

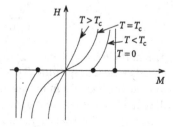

Let us now introduce the most important critical exponents. When doing this it is absolutely necessary to indicate precisely the path on which the change of state is brought about (Figs. 4.19 and 4.20):

1. $\boxed{\alpha, \alpha' : \textbf{Heat capacities}}$

For the **real gas** one agrees upon:

$$C_V \sim \begin{cases} A'\,(-\varepsilon)^{-\alpha'} & \left[\text{way II}, T \overset{\scriptscriptstyle <}{\to} T_c, \rho = \rho_{g,1}\right], \\ A\,\varepsilon^{-\alpha} & \left[\text{way I}, T \overset{\scriptscriptstyle >}{\to} T_c, \rho = \rho_c\right]. \end{cases} \tag{4.50}$$

The analog definition for the **magnet** reads:

$$C_H \sim \begin{cases} A'\,(-\varepsilon)^{-\alpha'} & \left[T \overset{\scriptscriptstyle <}{\to} T_C, \quad H = 0\right], \\ A\,\varepsilon^{-\alpha} & \left[T \overset{\scriptscriptstyle >}{\to} T_C, \quad H = 0\right]. \end{cases} \tag{4.51}$$

The experiment yields $\alpha, \alpha' \approx 0$. The exact (!) solution of the two-dimensional Ising model comes to a logarithmic divergence of the heat capacity C_V, i.e. to $\alpha = \alpha_S = 0$. The so-called **classical theories** (Weiss ferromagnet, van der Waals gas, Landau theory, ..., see Exercise 4.3.10) exhibit discontinuities, i.e. $\alpha = 0$.

2. $\boxed{\beta : \textbf{Order parameter}}$

The *order parameter* is understood as a variable which is meaningful **only in one** of the two phases which participate in the transition, or being unequal zero only in one of the two phases. The appearance of the order parameter thus signalizes the phase transition.

The order parameter of the magnet is the spontaneous magnetization $M_S(T)$, which appears only below T_C. For the real gas it is the density-difference $\Delta\rho = \rho_l - \rho_g$ or $\rho_{l,g} - \rho_c$, in the two-phase region. The critical behavior of the order parameter is described by the exponent β:

$$\frac{\Delta\rho(T)}{2\rho_c} \sim B(-\varepsilon)^\beta \qquad \text{(way II)}, \qquad (4.52)$$

$$\frac{M_S(T)}{M_S(0)} \sim B(-\varepsilon)^\beta \qquad (H = 0). \qquad (4.53)$$

The normalization factors $2\rho_c$ and $M_S(0)$, respectively, take care for that the so-called **critical amplitude** B has the order of magnitude of about 1 and does not vary too much from system to system.—Actually we should write β' instead of β since the exponent belongs to the low-temperature phase. But because the order parameter makes sense, by definition, only in one of the two phases, the distinction between β and β' appears to be unnecessary.

Typical experimental values for β are 0.35 ± 0.02. The classical theories yield all $\beta = 1/2$. For the $d = 2$-Ising model one finds exactly $\beta = 1/8$. For the $d = 3$-Ising model ($n = 1$) one gets by reliable approaches $\beta = 0.325 \pm 0.001$, for the $d = 3$-XY-model ($n = 2$) $\beta = 0.345 \pm 0.002$, and for the $d = 3$-Heisenberg model $\beta = 0.365 \pm 0.002$.

3. $\boxed{\gamma, \gamma' : \textbf{Compressibilities, Susceptibilities}}$

Because of

$$\kappa_T = -\frac{1}{V}\left(\frac{\partial V}{\partial p}\right)_T = \frac{1}{\rho}\left(\frac{\partial \rho}{\partial p}\right)_T,$$

$$\chi_T = \left(\frac{\partial M}{\partial H}\right)_T \qquad (4.54)$$

κ_T^{-1} and χ_T^{-1} correspond to the slopes of the isotherms in the $p - \rho$ diagram and the $H - M$ diagram, respectively. κ_T and χ_T will therefore diverge for $T \to T_c$. By convention, however, one chooses the critical exponents to be always positive:

$$\frac{\kappa_T}{\kappa_{T_c}^{(0)}} \sim \begin{cases} C'(-\varepsilon)^{-\gamma'} & \left[\text{way II, } T \xrightarrow{\;\;\;} T_c, \rho = \rho_{g,1}\right], \\ C\varepsilon^{-\gamma} & \left[\text{way I, } T \xrightarrow{\;\;\;} T_c, \rho = \rho_c\right]. \end{cases} \qquad (4.55)$$

$\kappa_{T_c}^{(0)}$ is the compressibility of the ideal gas at $T = T_c$:

$$\kappa_T^{(0)} = \frac{1}{p} = \frac{V}{nRT}.$$

Analogously to this, one uses for the magnetic system as normalization factor the susceptibility of the ideal paramagnet for which we have according to (1.25)

$$\chi_T^{(0)} = \frac{C^*}{T} \, ,$$

where C^* is the Curie constant defined in (1.26):

$$\frac{\chi_T}{\chi_{T_c}^{(0)}} \sim \begin{cases} C' \, (-\varepsilon)^{-\gamma'} & \left[T \overset{<}{\to} T_C, H = 0\right] \, , \\ C \, \varepsilon^{-\gamma} & \left[T \overset{>}{\to} T_C, H = 0\right] \, . \end{cases} \tag{4.56}$$

The experimental values for γ and γ' wobble a bit. The different measuring methods deliver somewhat different values around 1.3 with $\gamma \approx \gamma'$. The model calculations all give $\gamma = \gamma'$, where the classical theories all yield $\gamma = 1$. The $d = 2$-Ising model leads to $\gamma = 7/4$, refined approaches to the $d = 3$-Ising model find $\gamma \approx 1.24$, the $d = 3$-XY model gives $\gamma \approx 1.32$, and the $d = 3$-Heisenberg model $\gamma \approx 1.39$. The numerical values of the model calculations show nicely, as already seen also for α and β, the dependence of the critical exponents on the spin and lattice dimension.

4. $\boxed{\delta : \textbf{Critical isotherm}}$

When $p_c^{(0)} = k_B \, T_c \, \rho_c$ means the pressure of the ideal gas at $\rho = \rho_c$ and $T = T_c$, then for the real gas we have:

$$\frac{(p - p_c)}{p_c^{(0)}} \sim D \left| \frac{\rho}{\rho_c} - 1 \right|^\delta \, \text{sign} \, (\rho - \rho_c) \quad [\text{way III} , \, T = T_c] \, . \tag{4.57}$$

Here sign $(\rho - \rho_c)$ denotes the sign of $(\rho - \rho_c)$:

$$\text{sign} \, (\rho - \rho_c) = \frac{\rho - \rho_c}{|\rho - \rho_c|} \, .$$

δ thus corresponds more or less to the power of the function (polynomial) of the critical isotherm. The larger is δ the flatter is the isotherm.

If one takes

$$H_C^{(0)} = \frac{k_B \, T_c}{\mu_0 \, m} \qquad (m = \text{magnetic moment per particle}) \, ,$$

then the relation for the magnets, analogously to (4.57), reads:

$$\frac{H}{H_C^{(0)}} \sim D \left| \frac{M(T = T_c, H)}{M(T = 0, H = 0)} \right|^\delta \, \text{sign} \, (M) \, . \tag{4.58}$$

Experimental values for δ lie between 4 and 5. The result of the $d = 2$-Ising model $\delta = 15$ is distinctly out of line. For all the classical theories it is found $\delta = 3$. To the $d = 3$-Ising model, the $d = 3$-XY model and also the $d = 3$-Heisenberg model one ascribes $\delta \approx 4.8$. δ seems therefore to be influenced above all by the lattice dimension but not so much by the spin dimension.

Besides the critical exponents introduced from 1. to 4. there are still the exponents ν, ν' and η which are of particular importance. These are defined in connection with the pair-correlation function which we will get to know only in Statistical Mechanics. Therefore we leave open, at this stage, the exponents ν, ν' and η.

It is clear that the normalization factors $\kappa_{T_C}^{(0)}, \chi_{T_C}^{(0)}, p_c^{(0)}, H_C^{(0)}$ in the above definition equations are not of special importance. Very often they are thus left out. They only take care that the various quantities become dimension-less and that the amplitudes take the order of magnitude of 1.

Although, for instance, the critical temperature T_c strongly differs from material to material, we nevertheless recognize an astonishing similarity of the numerical values of the critical exponents.

4.2.4   Critical Exponent Inequalities

The theory of the critical exponents is at first based on pure hypotheses which get, however, strong support from experiment. Since, on the other hand, only very few really exact evaluations of realistic models are available, considerations are of course of great interest which lead, somehow, to the possibility of testing the theory. In this sense, some thermodynamically exact inequalities among the critical exponents have gained great importance. The most weighty ones are reviewed in this section where we concentrate ourselves here on the magnetic system as example.

For the following proofs we will frequently use the almost self-evident **lemma**:

If $f(x) \sim x^\varphi$ and $g(x) \sim x^\psi$, and furthermore for sufficiently small $|x|$ $|f(x)| \le |g(x)|$, then it must be

$$\varphi \ge \psi \tag{4.59}$$

Proof From $|f(x)| \le |g(x)|$ it follows

$$\ln |f(x)| \le \ln |g(x)|$$

and therewith for $|x| < 1$, i.e. $\ln |x| < 0$:

$$\frac{\ln |f(x)|}{\ln |x|} \ge \frac{\ln |g(x)|}{\ln |x|} .$$

According to (4.45) this is equivalent to the assertion (4.59).

We prove with this lemma at first the

Rushbrook-inequality

$$\alpha' + 2\beta + \gamma' \geq 2 \quad \text{for} \quad H = 0 , \quad T \to T_c^{(-)} . \tag{4.60}$$

Proof Starting point is the relation (2.82)

$$\chi_T(C_H - C_m) = \mu_0 V T \beta_H^2 = \mu_0 V T \left[\left(\frac{\partial M}{\partial T} \right)_H \right]^2 .$$

Because of $C_m \geq 0$, this leads to the inequality:

$$C_H \geq \mu_0 T V \left[\left(\frac{\partial M}{\partial T} \right)_H \right]^2 \chi_T^{-1} . \tag{4.61}$$

During the limiting process $\varepsilon \to 0$ the prefactor $\mu_0 T V$ remains finite and is therefore not essential. Because of

$$C_H \sim (-\varepsilon)^{-\alpha'}; \quad \chi_T \sim (-\varepsilon)^{-\gamma'} ; \quad M \sim (-\varepsilon)^\beta$$

it follows then with the above lemma (4.59):

$$-\alpha' \leq 2(\beta - 1) + \gamma' .$$

That agrees with the assertion (4.60).

There is some evidence that the Rushbrooke-inequality (4.60) can even be read as equality. Experimental results indicate that. For the classical theories ($\alpha' = 0$, $\beta = 1/2$, $\gamma' = 1$) the equality sign is exact. That holds also for the exactly solvable $d = 2$-Ising model ($\alpha' = 0$, $\beta = 1/8$, $\gamma' \approx 7/4$). Furthermore, it is approximately, but very trustworthily confirmed by the not exactly solved $d = 3$-Ising model. The scaling hypothesis, to be discussed in Sect. 4.2.5, takes the equality sign in (4.60).

It can be shown that the equal sign in (4.60) is valid if

$$R = \lim_{\varepsilon \to 0} \frac{C_m}{C_H} < 1 . \tag{4.62}$$

Note that $R \leq 1$ must always hold because of $C_m \leq C_H$. Let us prove the assertion (4.62):

1. $R = 1$

In this case, in the critical region, we should have

$$\frac{C_m}{C_H} \sim 1 - (-\varepsilon)^x (1 + \ldots) .$$

with $x > 0$. The minus sign guarantees $C_m \leq C_H$. We use again (2.82):

$$1 - \frac{C_m}{C_H} = \mu_0 V T \beta_H^2 \chi_T^{-1} C_H^{-1} . \qquad (4.63)$$

In the critical region this equation can be read as follows:

$$(-\varepsilon)^{x}(1 + \ldots) \sim (-\varepsilon)^{2(\beta-1)+\gamma'+\alpha'} (1 + \ldots) .$$

This has the consequence

$$x = 2(\beta - 1) + \gamma' + \alpha' ,$$

so that because of $x > 0$

$$\alpha' + 2\beta + \gamma' = 2 + x > 2$$

must be concluded. The Rushbrooke-relation is thus for $R = 1$ a true inequality.

2. $R = 1 - y < 1 \, (y > 0)$

The most general ansatz for the critical region is now:

$$\frac{C_m}{C_H} = 1 - y(1 + \varepsilon^x + \ldots) ; \quad x > 0 .$$

We insert this into (4.63):

$$1 - [1 - y(1 + \varepsilon^x + \ldots)] \sim (-\varepsilon)^{2(\beta-1)+\gamma'+\alpha'} .$$

The left-hand side remains finite and unequal zero for $\varepsilon \to 0$. This is, however, possible only if the exponent on the right-hand side is zero:

$$2 = 2\beta + \gamma' + \alpha' .$$

But this is just the Rushbrooke-relation (4.60) with the equality sign.

Next we derive the
Coopersmith-inequality

$$\varphi + 2\psi - \frac{1}{\delta} \geq 1 \quad \text{for} \quad T = T_C, \quad H \to 0^+ \qquad (4.64)$$

We have to respect the presumption $H \to 0^+$. Hence, the variable here is not ε, but H. The critical exponents φ and ψ are still unknown to us:

$$C_H \sim H^{-\varphi} ; \quad S(T_C, H) \sim -H^{\psi} \quad [T = T_C] . \qquad (4.65)$$

For the proof we use (4.58):

$$H \sim |M|^{\delta} \operatorname{sign} M \quad \Longrightarrow \quad M_+ \sim H^{1/\delta} .$$

That is inserted, together with the Maxwell relation

$$V \left(\frac{\partial M}{\partial T} \right)_{B_0} = \frac{1}{\mu_0} \left(\frac{\partial S}{\partial H} \right)_T \xrightarrow{T=T_C} \sim -H^{\psi-1}$$

into (4.61). For this purpose we still need the isothermal susceptibility χ_T, for which we cannot use here the exponents γ and γ', because these are defined for another way of the change of state:

$$\chi_{T_C} = \left(\frac{\partial M}{\partial H} \right)_{T_C} \sim H^{\frac{1}{\delta}-1} .$$

For $T = T_C$ and $H \to 0^+$ equation (4.61) can thus be written as follows:

$$H^{-\varphi}(1 + \ldots) \geq \frac{1}{\mu_0 V} T_C H^{2\psi-2} H^{1-\frac{1}{\delta}} (1 + \ldots) .$$

With the above-proven lemma (4.59) we come to

$$-\varphi \leq 2\psi - 2 + 1 - \frac{1}{\delta} ,$$

which directly leads to the assertion (4.64).

We want to finally derive a third important exponent inequality, namely the so-called
Griffiths-inequality

$$\alpha' + \beta(1 + \delta) \geq 2 \quad \text{for} \quad H = 0 , \quad T \to T_C^{(-)} . \tag{4.66}$$

According to the presumption the system is in the zero-field. For $T = T_1 \leq T_C$ we therefore denote by $M_1 = M_1(T_1)$ the **spontaneous** magnetization. Let M_0 be the saturation magnetization. For the free energy, in the ferromagnetic phase, we write (Fig. 4.21):

$$F(T_1, M) = F(T_1, 0) , \tag{4.67}$$

if $M < M_1(T_1)$. This means for the first derivative with respect to M:

$$\left(\frac{\partial F}{\partial M} \right)_{T_1} = \mu_0 V H = 0 ,$$

Fig. 4.21 Free energy of a ferromagnet as function of the magnetization M (*above*; M_1: spontaneous magnetization). Equation of state H–M of the ferromagnet (*below*)

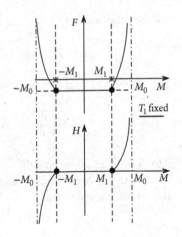

if $M < M_1(T_1)$. Using the Maxwell relation

$$\left(\frac{\partial S}{\partial M}\right)_T (T_1, M) = -\mu_0 V \left(\frac{\partial H}{\partial T}\right)_M (T_1, M) = 0 \quad M < M_1(T_1),$$

for the entropy we find the expression:

$$S(T_1, M) = S(T_1, 0), \quad \text{if} \quad M < M_1(T_1). \tag{4.68}$$

Let us define two new functions:

$$f(T, M) = (F(T, M) - F(T_C, 0)) + (T - T_C)S(T_C, 0), \tag{4.69}$$

$$s(T, M) = S(T, M) - S(T_C, 0). \tag{4.70}$$

Because of

$$S = -\left(\frac{\partial F}{\partial T}\right)_M$$

it obviously holds also:

$$s = -\left(\frac{\partial f}{\partial T}\right)_M.$$

In Sect. 4.2.1 we have seen that $F(T, M)$ is a concave function of T. Since the second derivatives of F and f with respect to T are identical, we can conclude that $f(T, M)$ is also a concave function of T. This we utilize now. The equation of the tangent at the $f(T, M_1)$-curve at T_1 reads:

$$\hat{f}(T, M_1) = f(T_1, M_1) + (T - T_1) \left(\frac{\partial f}{\partial T} \right)_M (T_1, M_1)$$
$$= f(T_1, M_1) - (T - T_1) \, s(T_1, M_1) \; .$$

Since f is concave as function of T, we can further conclude:

$$f(T, M_1) \le \hat{f}(T, M_1) \qquad \forall T \; .$$

This means, especially for $T = T_C$:

$$f(T_C, M_1) \le f(T_1, M_1) - (T_C - T_1) \, s(T_1, M_1) \; .$$

We insert now (4.67) and (4.68) into this inequality:

$$f(T_C, M_1) \le f(T_1, 0) - (T_C - T_1) \, s(T_1, 0) \; .$$

In the next step we show that $f(T_1, 0) \le 0$. According to the definition (4.70) we have $s(T_C, 0) = 0$, so that $f(T, 0)$ has a horizontal tangent at $T = T_C$. It follows therewith, since f is concave:

$$f(T_1, 0) \le f(T_C, 0) = 0 \; .$$

The above inequality is therefore more than ever valid in the form:

$$f(T_C, M_1) \le - (T_C - T_1) \, s(T_1, 0) \; . \tag{4.71}$$

This is eventually the inequality which we want to use for the estimation of the exponents. Let us start with the left-hand side. If we consider M_1 as variable, then along the critical isotherm it holds:

$$H(M_1, T = T_C) \sim M_1^{\delta} \; .$$

Because of

$$H = \frac{1}{\mu_0 V} \left(\frac{\partial F}{\partial M} \right)_{T_C} = \frac{1}{\mu_0 V} \left(\frac{\partial f}{\partial M} \right)_{T_C}$$

we come to:

$$f(T_C, M_1) \sim M_1^{\delta+1} \; .$$

On the other hand, M_1, as spontaneous magnetization, is also the order parameter so that we can exploit

$$M_1 \sim (T_C - T_1)^\beta$$

with the result:

$$f(T_C, M_1) \sim (T_C - T_1)^{\beta(\delta+1)} \ . \tag{4.72}$$

We now estimate the right-hand side of (4.71), where we begin with the critical behavior of the heat capacity C_H:

$$C_H = T\left(\frac{\partial S}{\partial T}\right)_H \sim (T_C - T)^{-\alpha'} \qquad \left[H = 0\ ; \quad T \to T_C^{(-)}\right] \ .$$

Since T itself behaves *uncritically*, it must already be:

$$\left(\frac{\partial S}{\partial T}\right)_{H=0} \sim (T_C - T)^{-\alpha'} \qquad \left[T \to T_C^{(-)}\right] \ .$$

We need information about the entropy $S = S(T, M)$ at $M = 0$:

$$dS = \left(\frac{\partial S}{\partial T}\right)_M dT + \left(\frac{\partial S}{\partial M}\right)_T dM$$

$$\Longrightarrow \quad \left(\frac{\partial S}{\partial T}\right)_H = \left(\frac{\partial S}{\partial T}\right)_M + \left(\frac{\partial S}{\partial M}\right)_T \left(\frac{\partial M}{\partial T}\right)_H \ .$$

According to (4.68) we have

$$\left(\frac{\partial S}{\partial M}\right)_T (T_1, M) = 0 \ , \quad \text{if} \quad M < M_1(T_1) \ ,$$

so that it follows:

$$\left(\frac{\partial S}{\partial T}\right)_{M=0} = \left(\frac{\partial S}{\partial T}\right)_{H=0} \sim (T_C - T)^{-\alpha'} \qquad \left(T \to T_C^{(-)}\right) \ .$$

It is therefore:

$$-s(T_1, 0) = S(T_C, 0) - S(T_1, 0) \sim (T_C - T_1)^{-\alpha'+1} \ .$$

This means, eventually, for the right-hand side of (4.71):

$$-(T_C - T_1)\, s(T_1, 0) \sim (T_C - T_1)^{2-\alpha'} \ . \tag{4.73}$$

Equations (4.71) to (4.73) yield by applying the lemma (4.59):

$$\beta(\delta + 1) \geq 2 - \alpha' .$$

This proves the validity of the Griffiths inequality (4.66)!

The classical theories ($\alpha' = 0, \beta = 1/2, \delta = 3$) as well as the exactly solvable $d = 2$-Ising model ($\alpha' = 0, \beta = 1/8, \delta = 15$) let expect in (4.66) even the equal sign.

4.2.5 Scaling Hypothesis

The last section has left the question unanswered whether the exact exponent inequalities can perhaps be even read as equalities. We could already find a series of indications for the equal sign. A further strong support for this assumption comes from the scaling hypothesis which we are now going to discuss. This hypothesis consists of a very simple ansatz for the formal structure of a certain thermodynamic potential. This ansatz is so far not yet mathematically strictly proven, but appears, however, to be *plausible* in many respects. Nevertheless, it is still a *hypothesis*.

As to the formulation of the scaling hypothesis we first recall the term *homogeneous function*, which we came across already in Sect. 1.2 of Vol. 2:

$f(x)$ is **homogeneous of degree** m, if for each $\lambda \in \mathbb{R}$

$$f(\lambda x) = \lambda^m f(x); .\tag{4.74}$$

If such a function is known at one point $x_0 \neq 0$, then $f(x)$ is fixed everywhere. For each x there exists namely a unique λ_x with $x = \lambda_x x_0$ such that $f(x) = \lambda_x^m f(x_0)$. One says that $f(x)$ is connected to $f(x_0)$ via a simple change of scale ('*scale transformation*').

We now extend the term 'homogeneity' to functions of more than one variable:

Definition 4.2.5 One calls $f(x, y)$ a

generalized homogeneous function,

if it holds for each $\lambda \in \mathbb{R}$

$$f\left(\lambda^a x, \lambda^b y\right) = \lambda f(x, y) ,\tag{4.75}$$

where a and b may be arbitrary real numbers.

In this sense, $f(x, y) = x^2 + 3y^5$, e.g., is a generalized homogeneous function with $a = 1/2$ and $b = 1/5$.

For the example of the free enthalpy $G(T, B_0)$ of a magnetic system ($B_0 = \mu_0 H$) we are now going to formulate the scaling hypothesis. We are interested here only in those parts of $G(T, B_0)$ which are non-analytic at T_C. Let all the others, the *uncritical* terms, be separated:

$$G(T, B_0) \quad \longrightarrow \quad G(\varepsilon, B_0) ; \quad \varepsilon = \frac{T - T_C}{T_C} .$$

Scaling Hypothesis (*Postulate of Homogeneity*)
$G(\varepsilon, B_0)$ *is a generalized homogeneous function, i.e., it holds for each* $\lambda \in \mathbb{R}$:

$$\mathbf{G} \left(\lambda^{a_\varepsilon} \varepsilon, \lambda^{a_B} \mathbf{B_0} \right) = \lambda \, \mathbf{G}(\varepsilon, B_0) . \tag{4.76}$$

The numbers a_ε and a_B will **not** be further specified, so that the scaling hypothesis will not be able to provide concrete numerical values for the critical exponents. It leads, however, to various relations between the exponents. As already mentioned, the ansatz (4.76) cannot be proven mathematically exactly. It is made *very plausible*, though, by the so-called **Kadanoff construction** performed on the example of the Ising model. At this stage, however, we cannot go into those details. That is done in the Sect. 4.2.2 of Vol. 8.—The scaling hypothesis has been formulated here for the free enthalpy. It transfers of course in a definite manner to the other thermodynamic potentials, too.

We will now at first show that all critical exponents can be expressed by a_ε and a_B. This will mean that by the determination of two exponents all the others will already be fixed.

We differentiate (4.76) partially with respect to B_0:

$$\lambda^{a_B} \frac{\partial}{\partial \left(\lambda^{a_B} B_0 \right)} G \left(\lambda^{a_\varepsilon} \varepsilon, \lambda^{a_B} B_0 \right) = \lambda \frac{\partial}{\partial B_0} G(\varepsilon, B_0) .$$

We still have

$$\frac{\partial G}{\partial B_0} = -m = -V M ,$$

getting therewith:

$$\lambda^{a_B} M \left(\lambda^{a_\varepsilon} \varepsilon, \lambda^{a_B} B_0 \right) = \lambda \, M(\varepsilon, B_0) . \tag{4.77}$$

Out of this relation we will draw very far-reaching conclusions.

1. | **Exponent** β |

We use (4.77) for $B_0 = 0$:

$$M \left(\lambda^{a_\varepsilon} \varepsilon, 0 \right) = \lambda^{1 - a_B} M(\varepsilon, 0) .$$

That is correct for **each** λ, thus also for

$$\lambda = (-\varepsilon)^{-1/a_\varepsilon} .$$

It follows therewith:

$$M(\varepsilon, 0) = (-\varepsilon)^{\frac{1-a_B}{a_\varepsilon}} M(-1, 0) .$$

$M(-1, 0)$ is a constant number. For $\varepsilon \to 0^-$ we can therefore write:

$$M(\varepsilon, 0) \sim (-\varepsilon)^{\frac{1-a_B}{a_\varepsilon}} .$$

The comparison with (4.53) yields:

$$\beta = \frac{1 - a_B}{a_\varepsilon} . \tag{4.78}$$

(Because of $B_0 = \mu_0 H$ $B_0 = 0$ means of course also $H = 0$). The critical exponent β is therefore completely fixed by the constants a_B and a_ε.

2. Exponent δ

We now insert $\varepsilon = 0$ into (4.77),

$$M(0, B_0) = \lambda^{a_B - 1} M(0, \lambda^{a_B} B_0) ,$$

and choose especially:

$$\lambda = B_0^{-1/a_B} .$$

This yields

$$M(0, B_0) = B_0^{\frac{1-a_B}{a_B}} M(0, 1)$$

with an unimportant constant $M(0, 1)$. For $\varepsilon = 0$ and $B_0 \to 0^+$ we thus can write:

$$M(0, B_0) \sim B_0^{\frac{1-a_B}{a_B}} \quad \Longleftrightarrow \quad B_0 \sim M(0, B_0)^{\frac{a_B}{1-a_B}} .$$

The comparison with (4.58) leads to:

$$\delta = \frac{a_B}{1 - a_B} . \tag{4.79}$$

Via (4.78) and (4.79) a_ε and a_B are completely determined by β and δ:

$$a_B = \frac{\delta}{1+\delta} \; ; \quad a_\varepsilon = \frac{1}{\beta} \frac{1}{1+\delta} \; . \tag{4.80}$$

If we succeed to express the other exponents by a_ε and a_B, then this will finally lead to relations between the critical exponents.

3. $\boxed{\text{Exponents } \gamma, \gamma'}$

For the susceptibility χ_T we have to evaluate

$$\chi_T = \left(\frac{\partial M}{\partial H} \right)_T = \mu_0 \left(\frac{\partial M}{\partial B_0} \right)_T \; .$$

For this purpose we can apply in the critical region again (4.77) by differentiating it partially with respect to the field B_0:

$$\lambda^{2a_B} \frac{\partial}{\partial(\lambda^{a_B} B_0)} M \left(\lambda^{a_\varepsilon} \varepsilon, \, \lambda^{a_B} B_0 \right) = \lambda \frac{\partial}{\partial B_0} M(\varepsilon, B_0) \; .$$

This yields:

$$\lambda^{2a_B} \chi_T \left(\lambda^{a_\varepsilon} \varepsilon, \, \lambda^{a_B} B_0 \right) = \lambda \, \chi_T \left(\varepsilon, B_0 \right) \; .$$

We take $B_0 = 0$ and choose:

$$\lambda = (\pm\varepsilon)^{-1/a_\varepsilon} \; .$$

We get therewith:

$$\chi_T(\varepsilon, 0) = (\pm\varepsilon)^{-\frac{2a_B - 1}{a_\varepsilon}} \chi_T(\pm 1, 0) \; .$$

The constant $\chi_T(\pm 1, 0)$ is again unimportant in the critical region, but possibly different for $T \to T_C^{(-)}$ and $T \to T_C^{(+)}$ (*critical amplitudes*). The comparison with (4.56) leads now to:

$$\gamma = \gamma' = \frac{2a_B - 1}{a_\varepsilon} \; . \tag{4.81}$$

4. Exponents α, α'

The heat capacity $C_H = C_{B_0}$ requires the second derivative of the free enthalpy with respect to the temperature:

$$C_H = C_{B_0} = -T \left(\frac{\partial^2 G}{\partial T^2} \right)_{B_0} = -\frac{T}{T_C^2} \left(\frac{\partial^2 G}{\partial \varepsilon^2} \right)_{B_0} .$$

The scaling hypothesis (4.76) is differentiated twice with respect to ε:

$$\lambda^{2a_\varepsilon} \frac{\partial^2}{\partial (\lambda^{a_\varepsilon} \varepsilon)^2} G (\lambda^{a_\varepsilon} \varepsilon, \lambda^{a_B} B_0) = \lambda \frac{\partial^2}{\partial \varepsilon^2} G(\varepsilon, B_0) .$$

This means:

$$\lambda^{2a_\varepsilon} C_H (\lambda^{a_\varepsilon} \varepsilon, \lambda^{a_B} B_0) = \lambda C_H(\varepsilon, B_0) .$$

We choose now

$$B_0 = 0 \quad \text{and} \quad \lambda = (\pm \varepsilon)^{-1/a_\varepsilon}$$

and obtain therewith:

$$C_H(\varepsilon, 0) = (\pm \varepsilon)^{-\frac{2a_\varepsilon - 1}{a_\varepsilon}} C_H(\pm 1, 0) .$$

The comparison with (4.51) fixes the critical exponents α and α':

$$\alpha = \alpha' = \frac{2a_\varepsilon - 1}{a_\varepsilon} . \tag{4.82}$$

A typical result of the scaling hypothesis consists in the fact that for $T \to T_c^{(-)}$ and for $T \to T_c^{(+)}$ the same critical exponents are found. *Primed* and *unprimed* exponents are always identical ($\alpha = \alpha'$, $\gamma = \gamma'$).

A second important result of the scaling hypothesis predicts the thermodynamically exact inequalities of the last section to be pure equalities, which are then called the

'scaling laws' .

At the end let us analyze some typical examples a bit in detail.
We combine (4.80) with (4.82):

$$\alpha' = 2 - \frac{1}{a_\varepsilon} = 2 - \beta(1 + \delta) \quad \Longrightarrow \quad \alpha' + \beta(1 + \delta) = 2. \tag{4.83}$$

This corresponds to the Griffiths-relation (4.66).

When we combine the equations (4.80) and (4.81) we get a connection between β, γ', and δ:

$$\gamma' = 2\frac{a_B}{a_\varepsilon} - \frac{1}{a_\varepsilon} = 2\beta\delta - \beta(1+\delta) = \beta\delta - \beta$$

$$\implies \quad \gamma' = \beta(\delta - 1) . \tag{4.84}$$

The corresponding thermodynamically exact inequality is the
Widom-inequality

$$\gamma' \geq \beta(\delta - 1) . \tag{4.85}$$

If one finally further combines (4.83) and (4.84) then one gets a connection between α, β, and γ':

$$\alpha' + 2\beta + \gamma' = 2 . \tag{4.86}$$

This relation we got to know as thermodynamically exact Rushbrooke-inequality (4.60).

There still exists a set of further thermodynamically exact inequalities. All of them turn into true equations in consequence of the scaling hypothesis. As to this point, however, we have to refer the reader here to the specialist literature!

4.3 Exercises

Exercise 4.3.1

1. With the Clausius-Clapeyron equation calculate explicitly the vapor pressure curve under the simplifying assumption that, to a good approximation, the mole volume satisfies

$$v_g \gg v_l \; ; \; v_g \approx \frac{RT}{p} .$$

Furthermore, it is allowed to assume for the molar evaporation heat $Q_M \approx$ const.. The vapor (gaseous phase) thus behaves almost like an ideal gas.

2. Determine the thermal expansion coefficient along the line of coexistence:

$$\beta_{coex} = \frac{1}{V}\left(\frac{\partial V}{\partial T}\right)_{coex} .$$

You can use here also $v_g \gg v_l$.

Exercise 4.3.2 Two coexisting gaseous phases, i.e., being at equilibrium, obey the thermal equations of state

$$pV_1 = \alpha_1 T ; \quad pV_2 = \alpha_2 T \quad (\alpha_1 \neq \alpha_2 : \text{ constants})$$

and have identical heat capacities

$$C_p^{(1,2)}(T) \equiv C_p(T) .$$

1. Show that the entropies of both the phases exhibit the same temperature dependence:

$$S_i(T,p) = g_i(p) + f(T) \quad i = 1,2 .$$

   Determine $g_i(p)$!
2. Determine the slope of the curve of coexistence:

$$\frac{d}{dT} p_{\text{coex}}$$

3. Calculate explicitly $p_{\text{coex}} = p_{\text{coex}}(T)$ and show that the transformation heat (latent heat) is constant along the line of coexistence!

Exercise 4.3.3 If one brings a superconductor of the first kind into a magnetic field **H**, there appears the so-called Meissner-Ochsenfeld effect, i.e., except for the edge layer, which we ignore, the magnetic induction vanishes in the interior of the superconductor:

$$\mathbf{B}_0 = \mu_0(\mathbf{H} + \mathbf{M}) = 0 .$$

If, however, **H** exceeds a critical field strength, which depends on temperature, a phase transition into the normal-conducting state takes place. To a good approximation it holds:

$$H_c(T) = H_0 \left[1 - (1 - \alpha) \left(\frac{T}{T_c} \right)^2 - \alpha \left(\frac{T}{T_c} \right)^4 \right]$$

(T_c = transition temperature).

1. Calculate the transformation (latent) heat by the use of the Clausius-Clapeyron equation. The magnetization of the normal-conducting phase (M_n) can thereby be neglected compared to that of the superconducting phase (M_s).

2. Calculate the *'stabilization-energy'* ΔG of the superconductor:

$$\Delta G = G_s(T, H = 0) - G_n(T, H = 0)$$

(n: normal-conducting, s: superconducting). Use again $M_n \ll M_s$.

3. Calculate the entropy-difference

$$\Delta S = S_s(T) - S_n(T)$$

thereby exploiting part 2. Compare the result with that from part 1.

4. What follows from the third law of thermodynamics for

$$\left(\frac{dH_C}{dT}\right)_{T=0} ?$$

5. Calculate the difference $\Delta C = C_s - C_n$ of the heat capacities!

6. Classify the phase transition!

Exercise 4.3.4   A physical quantity f in the critical region behaves as

$$f(T) = a T \ln |T - T_c| + b T^2 .$$

Find the corresponding critical exponent!

Exercise 4.3.5   Show that for phase transitions of second order *in the Ehrenfest sense* only critical exponents $\varphi = 0$ are possible!

Exercise 4.3.6   Determine the critical exponents of

$$\begin{aligned}
1. \quad & f(T) = a T^{5/2} - b , \\
2. \quad & f(T) = a T^2 + c (T - T_c)^{-1} , \\
3. \quad & f(T) = a \sqrt{|T - T_c|} + d ,
\end{aligned}$$

a, b, c, d: constants.

Exercise 4.3.7   Let the ratio of the heat capacities of a magnetic system

$$R = \frac{C_m}{C_H}$$

be temperature-**independent**. Show that the equal sign in the Rushbrooke-relation

$$\alpha' + 2\beta + \gamma' \geq 2$$

is valid if and only if $R \neq 1$.

Exercise 4.3.8 From the scaling hypothesis, for a magnetic system, derive the following relation for the magnetization M:

$$\frac{M(\varepsilon, H)}{(\pm\varepsilon)^\beta} = M\left(\pm 1, (\pm\varepsilon)^{-\beta\delta} H\right) .$$

Do you see a possibility to check the scaling hypothesis experimentally by this equation?

Exercise 4.3.9 By dint of the scaling hypothesis prove the following exponent-equation:

$$\begin{aligned}
&1. \quad \gamma(\delta + 1) = (2 - \alpha)(\delta - 1) , \\
&2. \quad \delta = \frac{2 - \alpha + \gamma}{2 - \alpha - \gamma} .
\end{aligned}$$

Exercise 4.3.10 Calculate the critical exponents β, γ, γ', and δ of the van der Waals gas:

1. First show that the equation of state of the van der Waals gas can be written in the reduced quantities

$$p_r = \frac{p}{p_c} - 1 ; \quad V_r = \frac{V}{V_c} - 1 ; \quad \varepsilon = \frac{T}{T_c} - 1$$

as follows

$$p_r \left(2 + 7V_r + 8V_r^2 + 3V_r^3\right) = -3V_r^3 + 8\varepsilon\left(1 + 2V_r + V_r^2\right) .$$

2. How does the reduced volume V_r behave for $T \overset{<}{\to} T_c$ and $T \overset{>}{\to} T_c$?
3. Determine the critical exponent β!
4. Show that on the critical isotherm it holds:

$$p_r = -\frac{3}{2} V_r^3 \left(1 - \frac{7}{2} V_r + \ldots\right)$$

5. Determine the critical exponent δ!
6. Use the compressibility κ_T to derive the values of the critical exponents γ and γ'. What can be said about the critical amplitudes C and C'?

Exercise 4.3.11 Investigate the *critical behavior* of the isobaric thermal expansion coefficient,

$$\beta = \frac{1}{V}\left(\frac{\partial V}{\partial T}\right)_p ,$$

for the van der Waals gas.

Exercise 4.3.12   Discuss the critical behavior of the Weiss ferromagnet (Sect. 1.4.4).

1. Show that by use of the reduced quantities,

$$\widehat{M} = \frac{M}{M_0} \; ; \quad b = \frac{m B_0}{k_B T} \; ; \quad \varepsilon = \frac{T - T_C}{T_C}$$

(m: magnetic moment; $M_0 = \frac{N}{V} m$: saturation magnetization), the equation of state can be written as follows:

$$\widehat{M} = L\left(b + \frac{3\widehat{M}}{\varepsilon + 1}\right)$$

($L(x) = \coth x - \frac{1}{x}$: *Langevin function*).

2. Calculate the critical exponent β!
3. What is the value of the critical exponent δ?
4. Derive the critical exponents γ, γ' and determine the ratio C/C' of the critical amplitudes!

4.4   Self-Examination Questions

To Sect. 4.1

1. Define the term *phase*?
2. Describe the *method of Lagrange multipliers*!
3. What is valid at thermodynamic equilibrium of an **isolated** system, consisting of π phases and α different components, for temperature T, pressure p, and the chemical potentials μ?
4. What is the general equilibrium condition for a closed system with $p = $ const and $T = $ const?
5. If the closed system consists of several phases and different components, and that at $T = $ const and $p = $ const, what can then be said about the chemical potential of the components in the case of equilibrium?
6. State and interpret Gibbs phase rule using H_2O as an example!
7. What is the number of phases that can exist at the most in a system of α components?
8. Outline the derivation of the Clausius-Clapeyron equation!
9. What is typical for a phase transition of **first** order?
10. What does one understand by the *Maxwell construction*?

11. What is seen as the *unphysical* behavior of the van der Waals isotherms for $T < T_c$?
12. How do the isotherms of the pV-diagram behave in the region of coexistence of vapor and liquid?
13. Give reasons for the Maxwell construction!
14. What does one understand by *lever rule*?

To Sect. 4.2

1. For what purpose is the transformation (latent) heat needed at the phase transition in the gas-liquid system?
2. Define and characterize a phase transition of first order!
3. Under which preconditions is the Clausius-Clapeyron equation applicable?
4. When is a function $f(x)$ convex and concave, respectively?
5. What does one understand by stability conditions?
6. Discuss the connection between the stability of a thermodynamic system and the convexity properties of its potential $G(T,p)$ and $F(T, V)$, respectively?
7. How does a phase transition of first order manifest itself in the first partial derivative of the free energy with respect to the volume V?
8. How does the transformation heat ΔQ change when one approaches the critical point (T_c, p_c) along the line of coexistence?
9. How is, according to Ehrenfest, the order of a phase transition determined?
10. Which conditions does a phase transition fulfill in the *Ehrenfest sense*? State examples for such a transition of second order!
11. What do we understand by the *Ehrenfest equations*?
12. Which criticism on the Ehrenfest classification of phase transitions comes into your mind?
13. How does the heat capacity C_p behave at a phase transition of **first** order?
14. What are the characteristics of a continuous phase transition?
15. Why the phase transitions of second order are of special physical interest?
16. What is the general definition of a critical exponent?
17. Under which restrictions are the critical exponents *universal*?
18. Which *critical behavior* of a physical quantity corresponds to the exponent $\varphi = 0$?
19. Which critical exponent describes the behavior of the heat capacity? Which value does the exponent take at a phase transition in the *Ehrenfest sense* (discontinuity at $T = T_c$)?
20. What is the *order parameter* of a magnet? Which exponent describes its critical behavior?
21. List typical numerical values for the critical exponents α, β, γ, δ.
22. For the real gas and for the magnet sketch the *critical isotherm*.
23. Which behavior does the critical exponent δ describe?
24. Which critical exponents are connected by the Rushbrooke-inequality?

25. What supports the equality sign in the Rushbrooke-relation?
26. State the Coopersmith-inequality?
27. What is the Griffiths-inequality?
28. What is the definition of a *generalized homogeneous function*?
29. For a magnetic system formulate the scaling hypothesis!
30. To which physical statements does the scaling hypothesis lead?
31. Is it possible to derive actual numerical values for the critical exponents by the scaling hypothesis?
32. Which relations are denoted as *scaling laws*?
33. Which information does the scaling hypothesis give about the critical exponents for the transitions $T \to T_C^{(-)}$ and $T \to T_C^{(+)}$, respectively, e.g., for γ and γ'?

Appendix A
Solutions of the Exercises

Section 1.6

Solution 1.6.1

1.

$$A(x, y) = \cos x \sin y \quad \Longrightarrow \quad \left(\frac{\partial A}{\partial y}\right)_x = \cos x \cos y \,,$$

$$B(x, y) = -\sin x \cos y \quad \Longrightarrow \quad \left(\frac{\partial B}{\partial x}\right)_y = -\cos x \cos y$$

$$\Longrightarrow \quad df \textbf{ not} \text{ a total differential .}$$

2.

$$A(x, y) = \sin x \cos y \quad \Longrightarrow \quad \left(\frac{\partial A}{\partial y}\right)_x = -\sin x \sin y \,,$$

$$B(x, y) = \cos x \sin y \quad \Longrightarrow \quad \left(\frac{\partial B}{\partial x}\right)_y = -\sin x \sin y$$

$$\Longrightarrow \quad \left(\frac{\partial A}{\partial y}\right)_x = \left(\frac{\partial B}{\partial x}\right)_y \quad \Longrightarrow \quad df \text{ total differential .}$$

© Springer International Publishing AG 2017
W. Nolting, *Theoretical Physics 5*, DOI 10.1007/978-3-319-47910-1

3.

$$A(x, y) = x^3 y^2 \implies \left(\frac{\partial A}{\partial y}\right)_x = 2x^3 y \,,$$

$$B(x, y) = -y^3 x^2 \implies \left(\frac{\partial B}{\partial x}\right)_y = -2y^3 x$$

$$\implies \left(\frac{\partial A}{\partial y}\right)_x \neq \left(\frac{\partial B}{\partial x}\right)_y \implies df \text{ \textbf{not} a total differential}.$$

Solution 1.6.2 One method of solution utilizes the Jacobian determinant (see exercise 1.7.1, Vol. 1). We choose here an alternative way.

We solve the functional relation for x and y:

$$x = x(y, z) \,; \quad y = y(x, z)$$

$$\implies dx = \left(\frac{\partial x}{\partial y}\right)_z dy + \left(\frac{\partial x}{\partial z}\right)_y dz \,,$$

$$dy = \left(\frac{\partial y}{\partial x}\right)_z dx + \left(\frac{\partial y}{\partial z}\right)_x dz \,.$$

Combining:

$$dx = \left(\frac{\partial x}{\partial y}\right)_z \left\{ \left(\frac{\partial y}{\partial x}\right)_z dx + \left(\frac{\partial y}{\partial z}\right)_x dz \right\} + \left(\frac{\partial x}{\partial z}\right)_y dz$$

$$\implies dx \left\{ 1 - \left(\frac{\partial x}{\partial y}\right)_z \left(\frac{\partial y}{\partial x}\right)_z \right\} = dz \left\{ \left(\frac{\partial x}{\partial y}\right)_z \left(\frac{\partial y}{\partial z}\right)_x + \left(\frac{\partial x}{\partial z}\right)_y \right\} \,.$$

Two variables are freely selectable $\implies dx, dz$ arbitrary $\implies$ coefficients must vanish:

$$1 - \left(\frac{\partial x}{\partial y}\right)_z \left(\frac{\partial y}{\partial x}\right)_z = 0 \,,$$

$$\left(\frac{\partial x}{\partial y}\right)_z \left(\frac{\partial y}{\partial z}\right)_x + \left(\frac{\partial x}{\partial z}\right)_y = 0 \,.$$

It follows from that:

1.

$$\left(\frac{\partial x}{\partial y}\right)_z = \frac{1}{\left(\frac{\partial y}{\partial x}\right)_z} \,.$$

2.

$$\left(\frac{\partial x}{\partial y}\right)_z \left(\frac{\partial y}{\partial z}\right)_x = -\left(\frac{\partial x}{\partial z}\right)_y \overset{1.}{=} -\frac{1}{\left(\frac{\partial z}{\partial x}\right)_y}$$

$$\implies \left(\frac{\partial x}{\partial y}\right)_z \left(\frac{\partial y}{\partial z}\right)_x \left(\frac{\partial z}{\partial x}\right)_y = -1 \; .$$

Solution 1.6.3

1. Starting point is the two-dimensional path integral

$$I(C) = \int_{A(C)}^{B} \left\{ \alpha(x,y)\,dx + \beta(x,y)\,dy \right\} \; .$$

Choose

$$\mathbf{Z} = \alpha(x,y)\,\mathbf{e}_x + \beta(x,y)\,\mathbf{e}_y$$
$$d\mathbf{r} = dx\,\mathbf{e}_x + dy\,\mathbf{e}_y \; .$$

Therewith it is:

$$I(C) = \int_{A(C)}^{B} \mathbf{Z} \cdot d\mathbf{r} \; .$$

Let C_1 and C_2 be two arbitrary paths in the xy-plane between A and B. C_1 and $-C_2$ then build a closed path C in the xy-plane:

$$I(C_1) + I(-C_2) = I(C_1) - I(C_2) = \oint_C \mathbf{Z} \cdot d\mathbf{r} \; .$$

The right-hand side can be evaluated by the use of the Stokes theorem:

$$\oint_C \mathbf{Z} \cdot d\mathbf{r} = \int_{F_C} \mathrm{curl}(\mathbf{Z}) \cdot d\mathbf{f} = \int_{F_C} \left(\frac{\partial \beta}{\partial x} - \frac{\partial \alpha}{\partial y} \right) \mathbf{e}_z \cdot d\mathbf{f} \; .$$

Hence:

$$I(C_1) - I(C_2) = \int_{F_C} \left(\frac{\partial \beta}{\partial x} - \frac{\partial \alpha}{\partial y} \right) df \; .$$

This means:

$$\frac{\partial \beta}{\partial x} = \frac{\partial \alpha}{\partial y} \implies I(C_1) = I(C_2) \; .$$

Since C and therewith also F_C are arbitrary, it can further be concluded:

$$I(C_1) = I(C_2) \implies \frac{\partial \beta}{\partial x} = \frac{\partial \alpha}{\partial y} \; .$$

Combining we get the condition for the path-independence of the integral:

$$I(C_1) = I(C_2) \iff \frac{\partial \beta}{\partial x} = \frac{\partial \alpha}{\partial y} \; .$$

That is equivalent to saying

$$\mathbf{Z} \cdot d\mathbf{r} = \alpha(x, y)\, dx + \beta(x, y)\, dy \equiv dF$$

represents a total differential. To say it differently:

$$dF = \frac{\partial F}{\partial x}\, dx + \frac{\partial F}{\partial y}\, dy = \alpha(x, y)\, dx + \beta(x, y)\, dy$$

is a total differential if and only if

$$\frac{\partial \alpha}{\partial y} = \frac{\partial \beta}{\partial x}$$

2. One recognizes that

$$dF = \alpha(x, y)\, dx + \beta(x, y)\, dy$$

represents a total differential:

$$\frac{\partial \alpha}{\partial y} = 2y e^x = \frac{\partial \beta}{\partial x} \; .$$

Therewith the integral I_{AB} is path-independent. The problem formulation is therefore reasonable. dF is the total differential of

$$F(x, y) = y^2 e^x$$

so that it holds

$$I_{AB} = F(1, 1) - F(0, 0) = e \; .$$

3. Because of

$$\frac{\partial \alpha}{\partial y} = 2e^x \neq \frac{\partial \beta}{\partial x} = y^2 e^x$$

δF is now not a total differential. The integral I_{AB} is path-dependent. The problem formulation is thus not reasonable since the indication of the path is lacking.

Solution 1.6.4 We use the solution to Exercise 1.6.2:

$$\kappa_T = -\frac{1}{V} \left(\frac{\partial V}{\partial p} \right)_T = -\frac{1}{V} \left[\left(\frac{\partial p}{\partial V} \right)_T \right]^{-1} ,$$

$$\beta = \frac{1}{V} \left(\frac{\partial V}{\partial T} \right)_p = -\frac{1}{V} \frac{1}{\left(\frac{\partial T}{\partial p} \right)_V \left(\frac{\partial p}{\partial V} \right)_T}$$

$$= -\frac{1}{V} \frac{\left(\frac{\partial p}{\partial T} \right)_V}{\left(\frac{\partial p}{\partial V} \right)_T} = \kappa_T \left(\frac{\partial p}{\partial T} \right)_V .$$

Solution 1.6.5

$$\left(\frac{\partial T}{\partial p} \right)_V \quad = \quad -\left[\left(\frac{\partial p}{\partial V} \right)_T \left(\frac{\partial V}{\partial T} \right)_p \right]^{-1} = -\frac{\left(\frac{\partial V}{\partial p} \right)_T}{\left(\frac{\partial V}{\partial T} \right)_p} = \frac{V \kappa_T}{V \beta}$$

(chain rule)

$$\Longrightarrow \quad \left(\frac{\partial T}{\partial p} \right)_V \quad = \quad \frac{V}{nR} + \frac{ap}{nR}$$

$$\Longrightarrow \quad T \quad = \quad T(V,p) = \frac{Vp}{nR} + \frac{a}{2nR} p^2 + G(V) .$$

In addition it must hold:

$$\left(\frac{\partial T}{\partial V} \right)_p = \frac{1}{V\beta} = \frac{p}{nR} = \frac{p}{nR} + 0 + G'(V)$$

$$\Longrightarrow \quad G'(V) = 0 \quad \Longleftrightarrow \quad G(V) = \text{const} = T_0 .$$

Therewith the equation of state reads:

$$pV + \frac{1}{2} ap^2 = nR(T - T_0) .$$

Solution 1.6.6

1. The isotherm of the van der Waals equation of state

$$\left(p + a\frac{n^2}{V^2}\right)(V - nb) = nRT$$

has an inflection point at the critical point:

$$\left(\frac{\partial p}{\partial V}\right)_{T_c} = \left(\frac{\partial^2 p}{\partial V^2}\right)_{T_c} \overset{!}{=} 0 \; ;$$

$$p = \frac{nRT}{V - nb} - a\frac{n^2}{V^2} \; ,$$

$$\left(\frac{\partial p}{\partial V}\right)_T = -\frac{nRT}{(V - nb)^2} + 2a\frac{n^2}{V^3} \; ,$$

$$\left(\frac{\partial^2 p}{\partial V^2}\right)_T = \frac{2nRT}{(V - nb)^3} - 6a\frac{n^2}{V^4} \; .$$

This means:

$$2a\frac{n^2}{V_c^3} = \frac{nRT_c}{(V_c - nb)^2} \; ; \quad 6a\frac{n^2}{V_c^4} = \frac{2nRT_c}{(V_c - nb)^3} \; .$$

Division of the left side by the right side:

$$\frac{1}{3}V_c = \frac{1}{2}(V_c - nb) \quad \Longrightarrow \quad V_c = 3nb \; ,$$

$$nRT_c = 2a\frac{n^2}{27n^3b^3}4n^2b^2 \quad \Longrightarrow \quad RT_c = \frac{8a}{27b} \; .$$

p_c follows then directly from the van der Waals equation:

$$p_c = \frac{8an}{27b}\frac{1}{2nb} - \frac{a}{9b^2} \quad \Longrightarrow \quad p_c = \frac{a}{27b^2} \; .$$

Therewith the constants are fixed:

$$a = \frac{9R}{8n}T_c V_c \; ; \quad b = \frac{V_c}{3n} \; .$$

2. It follows from 1.:

$$p_c V_c = \frac{3}{8}nRT_c \; .$$

Division of the van der Waals equation by this expression:

$$\left(\pi + \frac{a n^2}{p_c V^2}\right)\left(v - \frac{n b}{V_c}\right) = \frac{8}{3} t ,$$

$$\frac{a n^2}{V^2 p_c} = \frac{27 n^2 b^2}{V^2} = \frac{3}{v^2}$$

$$\implies \left(\pi + \frac{3}{v^2}\right)\left(v - \frac{1}{3}\right) = \frac{8}{3} t :$$

Law of corresponding states.

3.

$$\kappa_T = -\frac{1}{V}\left[\left(\frac{\partial p}{\partial V}\right)_T\right]^{-1} .$$

According to 1.:

$$\left(\frac{\partial p}{\partial V}\right)_T = -\frac{n R T}{\left(V - \frac{1}{3} V_c\right)^2} + \frac{9}{4}\frac{R}{n} T_c V_c \frac{n^2}{V^3} .$$

At $V = V_c$:

$$\left(\frac{\partial p}{\partial V}\right)_{T,V=V_c} = -\frac{9}{4}\frac{n R T}{V_c^2} + \frac{9}{4}\frac{n R T_c}{V_c^2}$$

$$\implies \kappa_T (V = V_c) = \frac{4 V_c}{9 n R}\frac{1}{T - T_c} \quad \left(T \overset{>}{\to} T_c\right) .$$

The add on $T \overset{>}{\to} T_c$ is important since only then the presumption $V = V_c$ can be realized. κ_T diverges like $(T - T_c)^{-1}$. At the critical point an arbitrarily small pressure change suffices to transform a finite volume of gas into liquid (condensation!). Keywords: Phase transition of second order, universality hypothesis, critical exponent $\gamma = 1$ (Fig. A.1).

Fig. A.1

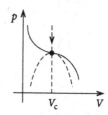

4. We have according to Exercise 1.6.4:

$$\beta = \kappa_T \left(\frac{\partial p}{\partial T} \right)_V \ ,$$

$$\left(\frac{\partial p}{\partial T} \right)_V = \frac{n R}{V - n b} \quad \Longrightarrow \quad \left(\frac{\partial p}{\partial T} \right)_{V = V_c} = \frac{3 \, n R}{2 \, V_c} \ .$$

β shows thus the same critical behavior as κ_T:

$$\beta = \beta \left(T, V = V_c \right) = \frac{2}{3} \frac{1}{T - T_c} \ .$$

Solution 1.6.7

1. Put

$$\bar{b} = \frac{b}{N_A} \ ; \quad \bar{a} = \frac{a}{k_B T N_A^2} \ .$$

Then it holds:

$$p = k_B T \rho \left(1 - \rho \bar{b} \right)^{-1} e^{-\bar{a}\rho}$$

$$= k_B T \rho \left(\sum_{\nu=0}^{\infty} \bar{b}^\nu \rho^\nu \right) \left(\sum_{\mu=0}^{\infty} \frac{1}{\mu!} (-\bar{a})^\mu \rho^\mu \right)$$

$$= k_B T \rho \left(1 + \sum_{n=1}^{\infty} B_n \rho^n \right) \ ,$$

$$B_n = \sum_{\mu=0}^{n} \frac{1}{\mu!} \bar{b}^{n-\mu} (-\bar{a})^\mu$$

$$\Longrightarrow \quad B_1 = \bar{b} - \bar{a} = \frac{1}{N_A} \left(b - \frac{a}{R T} \right) \ .$$

Boyle temperature T_B:

$$B_1 (T_B) \stackrel{!}{=} 0 \quad \Longrightarrow \quad T_B = \frac{a}{R b} \ .$$

2. Van der Waals:

$$p = N k_B T (V - n b)^{-1} - a \frac{n^2}{V^2}$$

$$= \rho k_B T \left(1 - \bar{b} \rho\right)^{-1} - k_B T \bar{a} \rho^2$$

$$= \rho k_B T \left(1 + \sum_{n=1}^{\infty} B_n \rho^n\right)$$

$$\implies \quad B_1 = \bar{b} - \bar{a} = \frac{1}{N_A} \left(b - \frac{a}{RT}\right) ; \quad B_n = \left(\frac{b}{N_A}\right)^n \quad \text{for} \quad n \geq 2 .$$

Dieterici:

B_1 as for the van der Waals model. Deviations appear not before B_2:

$$B_n = \sum_{\mu=0}^{n} \frac{1}{\mu!} \left(\frac{b}{N_A}\right)^{n-\mu} \left(-\frac{a}{k_B T N_A^2}\right)^{\mu} .$$

At high temperatures only the $\mu = 0$-term survives. Then one gets the same result as for the van der Waals model.

Meaning of the parameters:

Van der Waals:

$$a \rho^2 \sim intrinsic\ pressure,$$

$$b N \sim proper\ volume.$$

Dieterici:

$$v_{ij} : \quad interaction\ potential,$$

$$r : \quad distance.$$

a. b is, as in the van der Waals model, a measure of the proper volumes of the molecules, which are considered in the ideal gas as mathematical points. The particle repulsion gives rise to a *hard core*-potential (Fig. A.2).

Fig. A.2

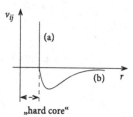

b. At larger distance an attraction of the particles sets in due to a mutual electric polarization of the atom shells, being connected to a tendency to a bound state, in which case one would no longer need a wall of the container. That means in any case a pressure decrease. This tendency is described in the Dieterici gas approximately by the exponential function (Fig. A.2):

$$a\,\rho \sim \text{average interaction energy}(\textit{activation energy}).$$

3. The equation of state

$$p = k_B\,T\,\rho\,(1 - \rho\,\bar{b})^{-1}\,e^{-\bar{a}\rho}$$

becomes of course *physically absurd* as soon as a particle densification ($\Longleftrightarrow$ increase of ρ) leads, because of the exponential function, to a pressure reduction. Sign-expectation:

$$\left(\frac{\partial p}{\partial \rho}\right)_T \geq 0 \,.$$

Because of $\kappa_T = -\frac{1}{V}\left(\frac{\partial V}{\partial p}\right)_T$ it is (N=const):

$$\left(\frac{\partial p}{\partial \rho}\right)_T = \left(\frac{\partial p}{\partial V}\right)_T \left(\frac{\partial V}{\partial \rho}\right)_T = \frac{-1}{V\,\kappa_T}\left[\left(\frac{\partial \rho}{\partial V}\right)_T\right]^{-1} = \frac{-1}{V\,\kappa_T}\left[-\frac{N}{V^2}\right]^{-1}$$

$$\Longrightarrow \quad \left(\frac{\partial p}{\partial \rho}\right)_T = \frac{1}{\rho\,\kappa_T}\,.$$

It is to require (*stability criterion*):

$$\kappa_T \geq 0 \,,$$

otherwise the system would collapse and therewith would be unstable.

4.

$$\left(\frac{\partial p}{\partial \rho}\right)_T = \frac{p}{\rho} + \frac{\bar{b}}{1 - \rho\,\bar{b}}\,p - \bar{a}\,p = p\left(\frac{1}{\rho\,(1 - \rho\,\bar{b})} - \bar{a}\right)$$

$$\Longrightarrow \quad \left(\frac{\partial p}{\partial \rho}\right)_{T_0} \overset{!}{=} 0 \quad \Longrightarrow \quad k_B\,T_0(\rho) = a^*\,\rho\,(1 - \rho\,\bar{b}) \qquad a^* = \frac{a}{N_A^2}$$

$$\Longrightarrow \quad k_B\,T_0(\rho) : \text{ parabola with zeros at } \rho = 0 \text{ and } \rho = \frac{1}{\bar{b}} \; (\textit{Fig. A.3})$$

Fig. A.3

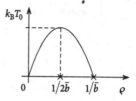

As *unphysical* one has to mark the region

$$\rho > \frac{1}{\bar{b}},$$

since then $T_0 < 0$ and $(\partial p / \partial \rho)_T < 0$, respectively. Approximately that means that $\rho > 1 / \bar{b}$, i.e. the proper volume of the molecules would be bigger than the total gas volume.

Maximum:

$$k_B \frac{dT_0}{d\rho} = a^* \left(1 - \rho \bar{b}\right) - a^* \bar{b} \rho = a^* \left(1 - 2 \rho \bar{b}\right) \overset{!}{=} 0 \implies \rho_C = \frac{1}{2\bar{b}},$$

$$k_B \left(\frac{d^2 T_0}{d\rho^2} \right)_{\rho = \rho_C} = -2 a^* \bar{b} < 0 \implies \text{maximum,}$$

$$k_B T_c = k_B T_0 \left(\rho_c\right) = \frac{a^*}{4 \bar{b}}.$$

The comparison with 1. yields:

$$T_B = 4 T_c.$$

The material constants a, b are determinable from the critical data!

$$a^* = \frac{a}{N_A^2} = \frac{2 k_B T_c}{\rho_c},$$

$$\bar{b} = \frac{b}{N_A} = \frac{1}{2 \rho_c}.$$

5. The region $\rho > 2 \rho_c \iff p < 0$ is unphysical.
   We investigate at first

$$\left(\frac{\partial p}{\partial \rho} \right)_T = p \left(\frac{1}{\rho(1 - \bar{b}\rho)} - \frac{a^*}{k_B T} \right)$$

with respect to zeros, for which it must obviously hold:

$$\rho(1 - \bar{b}\rho) \overset{!}{=} \frac{k_B T}{a^*} = \frac{1}{4\bar{b}} \frac{T}{T_c} .$$

This quadratic equation has two solutions:

$$\rho_{1,2} = \frac{1}{2\bar{b}} \left(1 \pm \sqrt{1 - \frac{T}{T_C}} \right) .$$

<u>$T > T_c$</u>

In this case $\left(\frac{\partial p}{\partial \rho} \right)_T$ has no real zero. Since otherwise

$$\left(\frac{\partial p}{\partial \rho} \right)_T (\rho \to 0) \to +\infty ,$$

it must be valid for all ρ:

$$\left(\frac{\partial p}{\partial \rho} \right)_T > 0 .$$

p is thus in any case a monotonically increasing function of ρ with $p(\rho = 0) = 0$.

<u>$T < T_c$</u> Now $\left(\frac{\partial p}{\partial \rho} \right)_T$ has two real(!) zeros at $\rho_{1,2}$.

$p(\rho)$ is for small ρ monotonically increasing. Hence the first zero corresponds to a maximum, the second to a minimum of $p(\rho)$. In between there is an *unphysical* region since $(\partial p / \partial \rho)_T < 0$.

<u>$T = T_c$</u>

$$\left(\frac{\partial p}{\partial \rho} \right)_{T_C} (\rho = \rho_c) = 0 ,$$

$$\left(\frac{\partial^2 p}{\partial \rho^2} \right)_T = p \left(\frac{1}{\rho(1 - \rho \bar{b})} - \frac{a^*}{k_B T} \right)^2 - p \frac{1 - 2\rho \bar{b}}{\rho^2 (1 - \rho \bar{b})^2} ,$$

$$\left(\frac{\partial^2 p}{\partial \rho^2} \right)_{T_C} (\rho = \rho_c) = 0 . \qquad \text{inflection point}$$

For $T < T_C$ a phase transition appears $\Longrightarrow$ *gas liquefaction!* The *unphysical* region is *corrected* by the same Maxwell construction ($A = B$!) as for the van der Waals gas (Fig. A.4).

Fig. A.4

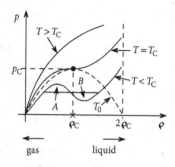

Solution 1.6.8

$$\kappa_T = -\frac{1}{V}\left(\frac{\partial V}{\partial p}\right)_T \; ; \quad \beta = \frac{1}{V}\left(\frac{\partial V}{\partial T}\right)_p ,$$

ideal gas: $p V = n R T$

$$\implies \quad \left(\frac{\partial V}{\partial p}\right)_T = -\frac{n R T}{p^2} = -\frac{V}{p} ,$$

$$\left(\frac{\partial V}{\partial T}\right)_p = \frac{n R}{p} = \frac{V}{T}$$

$$\implies \quad \kappa_T^{id} = \frac{1}{p} \; ; \quad \beta^{id} = \frac{1}{T} .$$

1.

$$p\,(V - n b) = n R T$$

$$\implies \quad \left(\frac{\partial V}{\partial p}\right)_T = -\frac{n R T}{p^2} = -\frac{1}{p}(V - n b) ,$$

$$\left(\frac{\partial V}{\partial T}\right)_p = \frac{n R}{p} = \frac{1}{T}(V - n b)$$

$$\implies \quad \kappa_T = \frac{1}{p}\left(1 - \frac{n b}{V}\right) = \kappa_T^{id}\left(1 - \frac{n b}{V}\right) ,$$

$$\beta = \frac{1}{T}\left(1 - \frac{n b}{V}\right) = \beta^{id}\left(1 - \frac{n b}{V}\right) .$$

2.

$$pV = nRT(1 + A_1(T)p)$$

$$\Longrightarrow \left(\frac{\partial V}{\partial p}\right)_T = -\frac{nRT}{p^2} = -\frac{V}{p} + \frac{nRTA_1(T)}{p}$$

$$\Longrightarrow \kappa_T = \kappa_T^{\mathrm{id}}\left(1 - \frac{nRTA_1(T)}{V}\right) ,$$

$$\left(\frac{\partial V}{\partial T}\right)_p = \frac{nR}{p} + nRA_1 + nRT\frac{dA_1}{dT} = \frac{V}{T} + nRT\frac{dA_1}{dT}$$

$$\Longrightarrow \beta = \beta^{\mathrm{id}}\left(1 + \frac{nRT^2}{V}\frac{dA_1}{dT}\right) .$$

3.

$$pV = nRT\left(1 + \frac{B_1(T)}{V}\right) ,$$

$$\left(\frac{\partial(pV)}{\partial p}\right)_T = V + p\left(\frac{\partial V}{\partial p}\right)_T = -\frac{nRTB_1}{V^2}\left(\frac{\partial V}{\partial p}\right)_T$$

$$\Longrightarrow \left(\frac{\partial V}{\partial p}\right)_T = -\frac{V}{p}\left(1 + \frac{nRTB_1}{pV^2}\right)^{-1}$$

$$\Longrightarrow \kappa_T = \kappa_T^{\mathrm{id}}\left(1 + \frac{nRTB_1(T)}{pV^2}\right)^{-1} ,$$

$$p\left(\frac{\partial V}{\partial T}\right)_p = nR\left(1 + \frac{B_1}{V}\right) + \frac{nRT}{V}\frac{dB_1}{dT} - \frac{nRTB_1}{V^2}\left(\frac{\partial V}{\partial T}\right)_p$$

$$\Longrightarrow \left(\frac{\partial V}{\partial T}\right)_p\left(p + \frac{nRTB_1}{V^2}\right) = \frac{pV}{T} + \frac{nRT}{V}\frac{dB_1}{dT}$$

$$\Longrightarrow \beta = \beta^{\mathrm{id}}\frac{1 + \frac{nRT^2}{pV^2}\frac{dB_1}{dT}}{1 + \frac{nRT}{pV^2}B_1} .$$

Solution 1.6.9

$$\delta W = B_0\,dm : \quad m : \quad \text{magnetic moment,}$$

$$dm = V\,dM : \quad M : \quad \text{magnetization, } V = \text{const}$$

$$\Longrightarrow \delta W = \mu_0 V H\,dM .$$

Fig. A.5

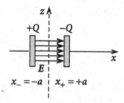

Curie law: $M = \frac{C}{T}H$

$$(\delta W)_T = \mu_0 \frac{C\,V}{T} H\,dH$$

$$\implies \quad \Delta W_{12} = \int_{H_1}^{H_2} (\delta W)_T = \frac{\mu_0\,C\,V}{T}\frac{1}{2}\left(H_2^2 - H_1^2\right) = \mu_0\,\frac{V\,T}{2\,C}\left(M_2^2 - M_1^2\right)\ .$$

Solution 1.6.10 According to equation (2.51), Vol. 3 it holds for the electric field inside the capacitor (Fig. A.5)

$$\mathbf{E} = \frac{Q}{\varepsilon_0 F_0}\,\mathbf{e}_x\ , \qquad F_0:\ \text{area of the plates.}$$

Capacity ((2.55), Vol. 3):

$$C = \varepsilon_0 \frac{F_0}{2\,a}\ .$$

Energy within the capacitor ((2.58), Vol. 3):

$$\overline{W} = \frac{1}{2}\frac{Q^2}{C} = \frac{Q^2}{2\,\varepsilon_0}\frac{2\,a}{F_0}\ .$$

Force on the plates of the capacitor (section 2.2.1, Vol. 3):

$$\mathbf{F}(+Q) = Q\,\mathbf{E}(x_-) = \frac{Q^2}{2\,\varepsilon_0 F_0}\,\mathbf{e}_x\ ,$$

$$\mathbf{F}(-Q) = -Q\,\mathbf{E}(x_+) = -\frac{Q^2}{2\,\varepsilon_0 F_0}\,\mathbf{e}_x\ .$$

Change of the capacity:

$$C = \frac{\varepsilon_0 F_0}{x_+ - x_-}\ .$$

Let x_- be variable:

$$\frac{dC}{dx_-} = \frac{\varepsilon_0 F_0}{(x_+ - x_-)^2} = \frac{C}{x_+ - x_-} \implies dx_- = \frac{dC}{C}(x_+ - x_-)$$

$dx_- > 0$ means distance-reduction leading to $dC > 0$. Analogously one finds dx_+.

$$dx_\pm = \mp \frac{dC}{C}(x_+ - x_-) \ .$$

1. Mechanical work due to distance-reduction:

$$\delta W = -\mathbf{F}(+Q)\, d\mathbf{x}_- = -\mathbf{F}(+Q)\, \mathbf{e}_x\, dx$$

$$= -\frac{Q^2}{2\,\varepsilon_0 F_0}\, \frac{(x_+ - x_-)}{C}\, dC = -\frac{1}{2}\frac{Q^2}{C^2}\, dC \ ,$$

$$\delta A = -\delta W = \frac{1}{2}\frac{Q^2}{C^2}\, dC \ .$$

That is the work done on the system *from the outside*.
2. Change of the field energy:

before:

$$\overline{W}_b = \frac{1}{2}\frac{Q^2}{C} \ ,$$

after:

$$\overline{W}_a = \frac{1}{2}\frac{Q^2}{C + dC} \ ,$$

$$d\overline{W} = \overline{W}_a - \overline{W}_b = \frac{Q^2}{2}\left(\frac{1}{C + dC} - \frac{1}{C}\right) \approx -\frac{1}{2}\frac{Q^2}{C^2}\, dC \ .$$

The change of the field energy thus corresponds to the work done *from the outside* on the system:

$$\delta A + \delta \overline{W} = 0 \ .$$

3. The last relation is always valid, i.e., for $dC > 0$ as well as for $dC < 0$. The state change is therefore reversible!

Solution 1.6.11

1. We have

$$\left(\frac{\partial B_0}{\partial m}\right)_T = \alpha T \left(\frac{1}{m_0 + m} + \frac{1}{m_0 - m}\right) - \gamma$$

$$= \alpha T \frac{2m_0}{m_0^2 - m^2} - \gamma = \frac{2\alpha}{m_0} T \left[1 - \frac{m^2}{m_0^2}\right]^{-1} - \gamma$$

$$= \frac{2\alpha}{m_0} \left(T \left[1 - \frac{m^2}{m_0^2}\right]^{-1} - T_C\right).$$

Instabilities follow therewith for

$$T \left[1 - \frac{m^2}{m_0^2}\right]^{-1} < T_C \iff |m| < m_0 \sqrt{1 - \frac{T}{T_C}}.$$

This means that for $T < T_C$ and $|m| < m_0 \sqrt{1 - \frac{T}{T_C}}$ the isotherms of the model-equation of state become unphysical. For $T > T_C$ and arbitrary $(-m_0 \le m \le +m_0)$ there are basically no instabilities. The isotherms are then *physical* in the whole allowed space.

2. Limiting curve:

$$m_S(T) = \pm m_0 \sqrt{1 - \frac{T}{T_C}}.$$

3. The ferromagnet is characterized by a *'spontaneous'* magnetic moment, which is **not** enforced by an external field. We have to therefore seek solutions of the equation

$$B_0(T, m) = 0.$$

One recognizes directly from the equation of state that the non-magnetic case $m = 0$ always represents a possible solution. That corresponds to paramagnetism. Of special interest is therefore the question whether there does exist an additional solution

$$m = m_S \ne 0.$$

(a) $\underline{T > T_C}$

$$\left(\frac{\partial B_0}{\partial m}\right)_T = \frac{2\alpha}{m_0}\left(T\left[1-\frac{m^2}{m_0^2}\right]^{-1} - T_C\right)$$

$$> \frac{2\alpha}{m_0}\left(T\left[1-\frac{m^2}{m_0^2}\right]^{-1} - T\right) > 0 .$$

$B_0 = B_0(T,m)$ is therewith for $T > T_C$ a bijective monotonic increasing function of m, and $m = 0$ is the only zero of $B_0(T,m)$ for $T > T_C$.

(b) $\underline{T < T_C}$

At first it is clear that even now $m = 0$ represents a possible solution. The equation of state shows in addition that, if $m_S \neq 0$ is a further solution, then $-m_S$ is also a zero. B_0-zeros have to fulfill the equation (Fig. A.6):

$$0 \overset{!}{=} \underbrace{\frac{T}{2T_C}\ln\left(\frac{m_0+m}{m_0-m}\right)}_{f(m)} - \underbrace{\frac{m}{m_0}}_{g(m)}$$

As to be seen in Fig. A.6 $g(m)$ is a straight line through the origin with the slope $1/m_0$ and the end points $g(\pm m_0) = \pm 1$. $f(m)$ has the properties $f(0) = 0$; $f(\pm m_0) = \pm\infty$. The slope is given by:

$$f'(m) = \frac{T}{2T_C}\left(\frac{1}{m_0+m} + \frac{1}{m_0-m}\right) \curvearrowright f'(0) = \frac{T}{T_C}\cdot\frac{1}{m_0} .$$

Because of $T < T_C$ the slope of f at the origin is smaller than that of g. Fig. A.6 illustrates that then two additional intersection points of f and g must exist and therefore zeros of $B_0(T,m)$ at $m = \pm m_S$.

4. Isotherms:

The very simple model already yields a qualitatively accurate description of the ferro-/paramagnet (Fig. A.7). However, below T_C we observe an unphysical behavior (see 1.). There the curves are to be replaced by a linear piece of the

Fig. A.6

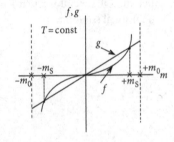

Fig. A.7

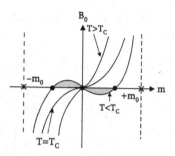

m-axis from $-m_S$ to $+m_S$, similar to the Maxwell construction for the van der Waals gas (Fig. 1.4).

Solution 2.9.1

1. First law of thermodynamics (gas!):

$$dU = \delta Q + \delta W = \delta Q - p\,dV\,,$$

$$U = U(T, V) \quad \Longrightarrow \quad dU = \left(\frac{\partial U}{\partial T}\right)_V dT + \left(\frac{\partial U}{\partial V}\right)_T dV$$

$$\Longrightarrow \quad \delta Q = \left(\frac{\partial U}{\partial T}\right)_V dT + \left[\left(\frac{\partial U}{\partial V}\right)_T + p\right] dV\,.$$

Integrability condition:

$$\left[\frac{\partial}{\partial V}\left(\frac{\partial U}{\partial T}\right)_V\right]_T \overset{!}{=} \left\{\frac{\partial}{\partial T}\left[\left(\frac{\partial U}{\partial V}\right)_T + p\right]\right\}_V$$

$$\Longrightarrow \quad \left[\frac{\partial}{\partial V}\left(\frac{\partial U}{\partial T}\right)_V\right]_T \overset{!}{=} \left[\frac{\partial}{\partial T}\left(\frac{\partial U}{\partial V}\right)_T\right]_V + \left(\frac{\partial p}{\partial T}\right)_V\,.$$

dU total, therefore

$$0 \overset{!}{=} \left(\frac{\partial p}{\partial T}\right)_V\,.$$

That is a contradiction, hence δQ **can not be a total differential!**

2. Ideal gas:

$$\left(\frac{\partial U}{\partial V}\right)_T = 0 \qquad \text{(Gay-Lussac)},$$

$$\left(\frac{\partial U}{\partial T}\right)_V = C_V = \text{const}$$

$$\Longrightarrow \quad \delta Q = C_V\,dT + p\,dV\,.$$

a) $\mu = \mu(T)$ to be chosen so that

$$dy = \mu\, \delta Q \quad \text{total differential}$$

$$\Longleftrightarrow \quad \left[\frac{\partial}{\partial V}(C_V \mu)\right]_T \stackrel{!}{=} \left[\frac{\partial}{\partial T}(\mu\, p)\right]_V$$

$$\Longleftrightarrow \quad 0 = \left[\frac{\partial}{\partial T}(\mu\, p)\right]_V = \frac{d\mu}{dT}\, p + \mu\left(\frac{\partial p}{\partial T}\right)_V$$

$$\Longleftrightarrow \quad \mu\, \frac{nR}{V} = -\frac{nRT}{V}\, \frac{d\mu}{dT}$$

$$\Longleftrightarrow \quad \mu + T\frac{d\mu}{dT} = d(\mu\, T) = 0 \quad \Longleftrightarrow \quad \mu\, T = \text{const}.$$

The constant is arbitrarily chosen equal to $1 \Longrightarrow$ *integrating factor*: $\mu(T) = 1/T$. We have therewith

$$dy = \frac{\delta Q}{T} = dS \quad \text{total differential.}$$

Entropy of the ideal gas:

$$S(T, V) - S(T_0, V_0) = C_V \int_{T_0}^{T} \frac{dT'}{T'} + \int_{V_0}^{V} \frac{p}{T}\, dV'$$

$$= C_V \ln \frac{T}{T_0} + nR \ln \frac{V}{V_0}.$$

Test:

$$\left(\frac{\partial S}{\partial T}\right)_V = \frac{C_V}{T}\,; \quad \left(\frac{\partial S}{\partial V}\right)_T = \frac{nR}{V} = \frac{p}{T}\,.$$

b) $\mu = \mu(V)$ to be chosen so that

$$dy = \mu\, \delta Q = (\mu\, C_V)\, dT + (\mu\, p)\, dV$$

becomes a total differential.

Integrability condition:

$$C_V \frac{d\mu}{dV} = \mu \left(\frac{\partial p}{\partial T}\right)_V = \mu \frac{nR}{V}$$

$$\Longleftrightarrow \quad \frac{C_V}{nR} \frac{d\mu}{\mu} = \frac{dV}{V} \; .$$

Ideal gas:

$$nR = C_p - C_V = C_V (\gamma - 1) \; ; \quad \gamma = \frac{C_p}{C_V}$$

$$\Longrightarrow \quad d \ln \mu = (\gamma - 1) \, d \ln V = d \ln V^{\gamma - 1}$$

$$\Longleftrightarrow \quad d \ln \left(\mu \, V^{1-\gamma}\right) = 0 \quad \Longleftrightarrow \quad \mu \, V^{1-\gamma} = \text{const} \; , \quad \text{e.g.} = 1$$

$$\Longrightarrow \quad \mu(V) = V^{\gamma - 1} \quad \Longrightarrow \quad dy = V^{\gamma - 1} \, \delta Q \; ;$$

$$dy = C_V \, V^{\gamma - 1} \left(dT + (\gamma - 1) \frac{T}{V} \, dV\right) \; .$$

Solution 2.9.2

$$\Delta W = - \int_1^2 p(V) \, dV = -\text{const} \int_1^2 \frac{dV}{V^n} \; , \quad (n \neq 1)$$

$$= \text{const} \left[\frac{1}{V_2^{n-1}} - \frac{1}{V_1^{n-1}}\right] \frac{1}{n - 1} \; ,$$

$$\text{const} = p_1 V_1^n = p_2 V_2^n$$

$$\Longrightarrow \quad \Delta W = \frac{1}{n - 1} (p_2 V_2 - p_1 V_1) \quad .$$

Ideal gas:

$$\Delta U = C_V (T_2 - T_1) = \frac{C_V}{N k_B} (p_2 V_2 - p_1 V_1)$$

$$\Longrightarrow \quad \frac{\Delta Q}{-\Delta W} = \frac{\Delta U - \Delta W}{-\Delta W} = 1 - \frac{\Delta U}{\Delta W} = 1 + (1 - n) \frac{C_V}{n k_B} = \text{const} \; .$$

Special case $n = 1$:

$$p V = \text{const} \quad \Longleftrightarrow \quad \text{isotherm of the ideal gas}$$

$$\Longrightarrow \quad \Delta U = 0 \Longrightarrow \quad \frac{\Delta Q}{-\Delta W} = 1 \,.$$

Solution 2.9.3 Integration:

$$\ln p' \Big|_{p_0}^{p} = a \ln V' \Big|_{V_0}^{V} \,.$$

This means:

$$p(V) = p_0 \left(\frac{V}{V_0} \right)^a \curvearrowright p V = \frac{p_0}{V_0^a} V^{a+1} \,.$$

With the equation of state of the ideal gas

$$\frac{pV}{T} = \frac{p_0 V_0}{T_0}$$

it follows

$$V = \frac{p_0}{p} \frac{V^{a+1}}{V_0^a} = \frac{V}{V_0} \frac{T_0}{T} \frac{V^{a+1}}{V_0^a} \curvearrowright V^{a+1} = V_0^{a+1} \frac{T}{T_0}$$

and therewith

$$V(T) = V_0 \left(\frac{T}{T_0} \right)^{\frac{1}{a+1}} \quad (V_0 = V(T_0)) \,.$$

First law of thermodynamics for closed systems:

$$\delta Q = dU + p\, dV = \left(\frac{\partial U}{\partial T} \right)_V dT + \left(\frac{\partial U}{\partial V} \right)_T dV + p\, dV \,.$$

Ideal gas, Gay-Lussac: $\left(\frac{\partial U}{\partial V} \right)_T = 0$:

$$\delta Q = C_V dT + p\, dV \,.$$

Heat capacity:

$$C_a = \left(\frac{\delta Q}{dT} \right)_a = C_V + p \left(\frac{\partial V}{\partial T} \right)_a = C_V + \frac{p}{a+1} \frac{V}{T} = C_V + \frac{nR}{a+1} \,.$$

Special cases:

$$a = 0 : \quad p(V) = p_0 = \text{const.} \qquad \text{isobaric}$$
$$C_a = C_V + nR = C_p$$
$$a \to \infty : \quad V(T) = V_0 = \text{const.} \qquad \text{isochoric}$$
$$C_a = C_V$$
$$a \to -1 : \quad pV = p_0 V_0 = \text{const.} \qquad \text{isothermal}$$
$$a \to -\gamma : \quad pV^\gamma = p_0 V_0^\gamma = \text{const.} \qquad \text{adiabatic}$$
$$C_a = C_V - \frac{nR}{\gamma - 1}$$
$$= C_V - C_V \frac{nR}{C_p - C_V}$$
$$= C_V - C_V = 0$$

Solution 2.9.4

1. According to (2.59) it is generally valid:

$$\left(\frac{\partial U}{\partial V} \right)_T = T \left(\frac{\partial p}{\partial T} \right)_V - p .$$

Therewith:

$$\left(\frac{\partial U}{\partial V} \right)_T = 0 \; \curvearrowright \; \left(\frac{\partial p}{\partial T} \right)_V = \frac{p}{T} .$$

This means:

$$p = p(T, V) = T f(V) .$$

Valid in particular for the ideal gas!

2.

$$\left(\frac{\partial U}{\partial V} \right)_T = bp = T \left(\frac{\partial p}{\partial T} \right)_V - p$$

$$\curvearrowright \; \frac{b}{V} f(T) = \frac{T}{V} f'(T) - \frac{1}{V} f(T)$$

$$\curvearrowright \; \frac{1}{V} (b + 1) f(T) = \frac{T}{V} f'(T)$$

$$\curvearrowright \; \frac{df}{f} = (b + 1) \frac{dT}{T}$$

$$\curvearrowright \ln f(T) = (1+b) \ln T + c$$

$$\curvearrowright f(T) \propto T^{1+b} \text{ or } f(T) = p_0 V_0 \left(\frac{T}{T_0}\right)^{1+b} .$$

Solution 2.9.5 Adiabatic means $\delta Q = 0$ and therewith $dS = 0$ (isentropic). For the isotherm it is $dT = 0$. If there were, as plotted in Fig. A.8, two intersection points A and B then it would hold obviously:

$$\oint_{ABA} \delta W = -\oint_{ABA} p \, dV \neq 0 .$$

First law of thermodynamics:

$$\delta W = dU - \delta Q = dU - T \, dS .$$

It follows therewith:

$$\oint_{ABA} \delta W = \oint_{ABA} (dU - T \, dS) = -\oint_{ABA} T \, dS \quad \text{(since } dU \text{ total differential)}$$

$$= -\int_{C_1} T \, dS - \int_{C_2} T \, dS = -\int_{C_1} T \, dS \quad \text{(since } C_2 \text{ adiabatic, isentropic)}$$

$$= -T^* \int_{C_1} dS \quad \text{(since } C_1 \text{ isotherm)}$$

$$= -T^* (S(A) - S(B)) \quad \text{(since } dS \text{ total differential)}.$$

A and B lie on the same isotropic, i.e., $S(A) = S(B)$. The integral thus vanishes:

$$\oint_{ABA} \delta W = 0 .$$

This conflicts with Fig. A.8. The figure must therefore be wrong. Isotherms and adiabatics never intersect twice!

Fig. A.8

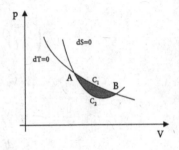

Solution 2.9.6

1.

$$C_m = \left(\frac{\partial U}{\partial T}\right)_m$$

follows directly from the first law, see (2.19).

We exploit (2.20):

$$C_H = C_m + \left[\left(\frac{\partial U}{\partial m}\right)_T - \mu_0 H\right]\left(\frac{\partial m}{\partial T}\right)_H .$$

Curie law:

$$m = V M = V \frac{C}{T} H$$

$$\implies \quad \left(\frac{\partial m}{\partial T}\right)_H = -V \frac{C}{T^2} H = -V \frac{M^2}{C H} .$$

Above inserted, this yields:

$$C_H = \left(\frac{\partial U}{\partial T}\right)_m + \left(\frac{\partial U}{\partial m}\right)_T \left(\frac{\partial m}{\partial T}\right)_H + \mu_0 \frac{V}{C} M^2 ,$$

$$U = U(T,m) \quad \implies \quad dU = \left(\frac{\partial U}{\partial T}\right)_m dT + \left(\frac{\partial U}{\partial m}\right)_T dm .$$

This means:

$$\left(\frac{\partial U}{\partial T}\right)_H = \left(\frac{\partial U}{\partial T}\right)_m + \left(\frac{\partial U}{\partial m}\right)_T \left(\frac{\partial m}{\partial T}\right)_H .$$

Hence we are left with the assertion:

$$C_H = \left(\frac{\partial U}{\partial T}\right)_H + \mu_0 \frac{V}{C} M^2 .$$

2. We can write:

$$\left(\frac{\partial m}{\partial H}\right)_{ad} = \left(\frac{\partial m}{\partial T}\right)_{ad} \left(\frac{\partial T}{\partial H}\right)_{ad} .$$

We determine the two factors separately:

a) First law of thermodynamics:

$$dU = \delta Q + \mu_0 H\, dm$$

$$= \left(\frac{\partial U}{\partial T}\right)_m dT + \left(\frac{\partial U}{\partial m}\right)_T dm\,,$$

$\delta Q = 0$, because the state change is adiabatic!

$$\Longrightarrow \quad \left\{\left(\frac{\partial U}{\partial T}\right)_m dT\right\}_{ad} = \left\{\left[\mu_0 H - \left(\frac{\partial U}{\partial m}\right)_T\right] dm\right\}_{ad}\,.$$

This means:

$$\left(\frac{\partial m}{\partial T}\right)_{ad} = \frac{C_m}{\mu_0 H - \left(\frac{\partial U}{\partial m}\right)_T}\,.$$

b) First law of thermodynamics:

$$0 = \delta Q = dU - \mu_0 H\, dm\,,$$

now: $U = U(T,H); \quad m = m(T,H)$.
   It follows:

$$0 = \left\{\left(\frac{\partial U}{\partial T}\right)_H - \mu_0 H \left(\frac{\partial m}{\partial T}\right)_H\right\} dT + \left\{\left(\frac{\partial U}{\partial H}\right)_T - \mu_0 H \left(\frac{\partial m}{\partial H}\right)_T\right\} dH\,.$$

According to part 1. the first bracket is equal to C_H:

$$C_H dT = \left\{\mu_0 H V \left(\frac{\partial M}{\partial H}\right)_T - \left(\frac{\partial U}{\partial H}\right)_T\right\} dH$$

$$= \left\{\mu_0 V M(T,H) - \left(\frac{\partial U}{\partial H}\right)_T\right\} dH\,.$$

From this it can be read off:

$$\left(\frac{dT}{dH}\right)_{ad} = \frac{\mu_0 m(T,H) - \left(\frac{\partial U}{\partial H}\right)_T}{C_H}\,.$$

The combination of a) and b) leads to the assertion:

$$\left(\frac{\partial m}{\partial H}\right)_{ad} = \left(\frac{\partial m}{\partial T}\right)_{ad} \left(\frac{\partial T}{\partial H}\right)_{ad} = \frac{C_m}{C_H} \frac{\mu_0 m - \left(\frac{\partial U}{\partial H}\right)_T}{\mu_0 H - \left(\frac{\partial U}{\partial m}\right)_T}\,.$$

Solution 2.9.7

1. The wall is thermally insulating, so the process running in the right chamber is adiabatic. By use of the adiabatic equations of the ideal gas (2.24), (2.25) we get ($\gamma = C_p / C_V$ given!):

$$\Delta W_r = -\int_{V_0}^{V_r} p \, dV = -\text{const}_1 \int_{V_0}^{V_r} \frac{dV}{V^\gamma} = -\frac{\text{const}_1}{1-\gamma} \left(V_r^{1-\gamma} - V_0^{1-\gamma} \right) ,$$

$$\text{const}_1 = p_0 V_0^\gamma = p_r V_r^\gamma$$

$$\implies \quad \Delta W_r = -\frac{1}{1-\gamma} (p_r V_r - p_0 V_0) = \frac{-N k_B}{1-\gamma} (T_r - T_0) .$$

We still fix T_r by the initial data:

$$T_r^\gamma p_r^{1-\gamma} = T_0^\gamma p_0^{1-\gamma} \quad \implies \quad T_r = T_0 \left(\frac{p_0}{p_r} \right)^{(1-\gamma)/\gamma}$$

$$\implies \quad \Delta W_r = -\frac{N k_B T_0}{1-\gamma} \left[\left(\frac{p_0}{p_r} \right)^{(1-\gamma)/\gamma} - 1 \right]$$

$$\implies \quad \Delta W_r = \frac{N k_B T_0}{\gamma - 1} \left(3^{(\gamma-1)/\gamma} - 1 \right) > 0 \quad \text{(because } \gamma > 1 \text{)} !$$

Hence work is done **on** the right system! Trivially: $\Delta Q_r = 0$.

2.

$$T_r^\gamma p_r^{1-\gamma} = T_0^\gamma p_0^{1-\gamma} ; \quad p_r = 3 p_0 \quad \implies \quad T_r = T_0 \, 3^{(\gamma-1)/\gamma} > T_0 .$$

Equation of state:

$$T_1 = \frac{p_1 V_1}{N k_B} = \frac{p_1}{N k_B} (2 V_0 - V_r) .$$

Equilibrium:

$$p_1 = p_r = 3 p_0$$

$$\implies \quad T_1 = \frac{3 p_0 \, 2 V_0}{N k_B} - T_r = 6 T_0 - T_r$$

$$\implies \quad T_1 = T_0 \left(6 - 3^{(\gamma-1)/\gamma} \right) .$$

3.

$$\text{First law:} \qquad \Delta Q_1 = \Delta U_1 - \Delta W_1 \, ,$$

$$\text{ideales Gas:} \qquad \Delta U_1 = C_V \, (T_1 - T_0) \, .$$

ΔW_1 takes care of the energy change on the right-hand side, i.e.:

$$\Delta W_1 = -\Delta W_r = -\Delta U_r \quad \text{(adiabatic on the right side)}$$

$$\Longrightarrow \quad \Delta W_1 = -C_V(T_r - T_0)$$

$$\Longrightarrow \quad \Delta Q_1 = C_V(T_1 - T_0 + T_r - T_0) = C_V(T_1 + T_r - 2\,T_0) \, ,$$

$$T_1 + T_r = 6\,T_0 \quad \Longrightarrow \quad \Delta Q_1 = 4\,C_V\,T_0 \, .$$

Solution 2.9.8

1.

$$\delta W = -p\,dV, \quad V = \frac{RT}{p}, \quad dV = -\frac{RT}{p^2}\,dp$$

$$\Longrightarrow \quad \Delta W = RT \int_{p_0}^{p_1} \frac{dp}{p} = RT\ln\frac{p_1}{p_0} = -RT\ln 20 < 0 \, .$$

The system therefore carries out work:

$$R = 8.315\,\frac{\text{J}}{\text{centigrade} \cdot \text{mole}} \quad \Longrightarrow \quad \Delta W = -7.298 \cdot 10^3 \, \text{J} \, .$$

2.

$$T = \text{const}\,, \quad \text{ideal gas} \quad \Longrightarrow \quad dU = 0$$

$$\Longrightarrow \quad \delta Q = -\delta W \quad \Longrightarrow \quad \Delta Q = |\Delta W| \, .$$

3. State change now adiabatic:

$$p\,V^\gamma = C \quad \Longrightarrow \quad dV = -\frac{1}{\gamma}\,C^{1/\gamma}\,\frac{1}{p^{1/\gamma+1}}\,dp$$

$$\Longrightarrow \quad \Delta W = \frac{C^{1/\gamma}}{\gamma} \int_{p_0}^{p_1} \frac{dp}{p^{1/\gamma}} = \frac{C^{1/\gamma}}{\gamma\left(1-\frac{1}{\gamma}\right)} \left[p_1^{1-1/\gamma} - p_0^{1-1/\gamma}\right] \, .$$

Now we have

$$C^{1/\gamma} p_{1,0}^{-1/\gamma} = V_{1,0}$$

$$\Longrightarrow \quad \Delta W = \frac{p_1 V_1 - p_0 V_0}{\gamma - 1},$$

$$p_1 V_1^{\gamma} = p_0 V_0^{\gamma} \Longrightarrow V_1 = V_0 \left(\frac{p_0}{p_1}\right)^{1/\gamma}$$

$$\Longrightarrow \quad \Delta W = \frac{V_0}{\gamma - 1}\left[p_1 \left(\frac{p_0}{p_1}\right)^{1/\gamma} - p_0\right] = \frac{R T_0}{\gamma - 1}\left[\left(\frac{p_0}{p_1}\right)^{1/\gamma - 1} - 1\right],$$

$$C_V = \frac{5}{2} R \Longrightarrow \gamma = \frac{C_p}{C_V} = 1 + \frac{R}{C_V} = 1.4,$$

$$T_0 = 293\,\mathrm{K} \quad ; \quad p_0 = 20\,p_1$$

$$\Longrightarrow \quad \Delta W = -3.503 \cdot 10^3\,\mathrm{J}.$$

4. Adiabatic process:

$$\Delta W = \Delta U = C_V (T_1 - T_0)$$

$$\Longrightarrow \quad T_1 = T_0 + \frac{\Delta W}{C_V} = 124.5\,\mathrm{K}.$$

Solution 2.9.9 A displacement by z means a volume change of

$$\Delta V = z F \qquad (F = \text{cross section area.})$$

By that a difference Δp between external and internal pressure arises which takes care for a repelling force in z-direction:

$$K = F \Delta p.$$

The state change of the ideal gas takes place adiabatically:

$$p V^{\gamma} = \text{const}.$$

By that we calculate Δp:

$$d(p V^{\gamma}) = 0 = dp\, V^{\gamma} + \gamma p V^{\gamma - 1}\, dV \Longrightarrow dp = -\frac{\gamma p}{V}\, dV.$$

This means

$$\Delta p = -\frac{\gamma p}{V}\, \Delta V = -\frac{\gamma p}{V}\, z F$$

and therewith

$$K = -\frac{\gamma p}{V} F^2 z = -k z \, .$$

If m is the mass of the sphere then the oscillation period reads:

$$\tau = 2\pi \sqrt{\frac{m}{k}} = 2\pi \sqrt{\frac{m V}{\gamma p F^2}} \quad \Longrightarrow \quad \gamma = \frac{4\pi^2 \, m \, V}{p \, F^2 \, \tau^2} \, .$$

Solution 2.9.10 First we determine

$$\frac{1}{T} \delta Q_{\text{rev}}$$

for both the systems as functions of T and V. For that first we use therefore the first law of thermodynamics,

$$\delta Q = dU + p \, dV \, ,$$

and the presumption,

$$U = U(T) \quad \Longrightarrow \quad dU = C_V(T) \, dT \, ,$$

for a reversible state change:

$$\frac{1}{T} \delta Q_{\text{rev}} = C_V(T) \frac{dT}{T} + \frac{p}{T} dV \, ,$$

$$p(A) = \alpha \frac{N T}{V^2} \, ; \quad p(B) = \left(\beta \frac{N}{V} T \right)^{1/2}$$

$$\Longrightarrow \quad \frac{1}{T} \delta Q_{\text{rev}} (A) = C_V(T) \frac{dT}{T} + \alpha N \frac{dV}{V^2} \, ,$$

$$\frac{1}{T} \delta Q_{\text{rev}} (B) = C_V(T) \frac{dT}{T} + \left(\beta \frac{N}{T V} \right)^{1/2} dV \, .$$

For the entropy to be a state quantity,

$$dS = \frac{1}{T} \delta Q_{\text{rev}}$$

must be a total differential. For that we check the integrability conditions:

(A)

$$\frac{\partial}{\partial V} \left(\frac{C_V(T)}{T} \right) \overset{!}{=} \frac{\partial}{\partial T} \left(\frac{\alpha N}{V^2} \right) \, ; \quad \text{obviously fulfilled!}$$

$$\Longrightarrow \quad \text{entropy as state quantity definable!}$$

(B)

$$\frac{\partial}{\partial V}\left(\frac{C_V(T)}{T}\right) \stackrel{!}{=} \frac{\partial}{\partial T}\left(\beta\frac{N}{VT}\right)^{1/2}$$

$$\Longleftrightarrow \quad 0 = -\frac{1}{2T}\left(\frac{\beta N}{TV}\right)^{1/2} .$$

This is a contradiction. An entropy is thus not definable for system B. System B therefore can not exist!

Solution 2.9.11

1. According to (2.59) it is:

$$\left[\left(\frac{\partial U}{\partial V}\right)_T + p\right] = T\left(\frac{\partial p}{\partial T}\right)_V .$$

This means according to (2.58):

$$dS = \frac{C_V(T,V)}{T}dT + \left(\frac{\partial p}{\partial T}\right)_V dV .$$

Integrability condition for dS:

$$\frac{1}{T}\left(\frac{\partial}{\partial V}C_V(T,V)\right)_T \stackrel{!}{=} \left(\frac{\partial^2 p}{\partial T^2}\right)_V = \left(\frac{\partial \alpha(V)}{\partial T}\right)_V = 0$$

$$\Longleftrightarrow \quad C_V(T,V) = C_V(T) .$$

2.

$$dS = \frac{C_V(T)}{T}dT + \left(\frac{\partial p}{\partial T}\right)_V dV .$$

The van der Waals gas fulfills the presumptions of part 1.:

$$\left(p + a\frac{n^2}{V^2}\right)(V - nb) = nRT$$

$$\Longrightarrow \quad p = T\frac{nR}{V - nb} - a\frac{n^2}{V^2} \quad \Longrightarrow \quad \left(\frac{\partial p}{\partial T}\right)_V = \frac{nR}{V - nb}$$

$$\Longrightarrow \quad dS = \frac{C_V}{T}dT + \frac{nR}{V - nb}dV$$

C_V according to the presumption T-independent

$$\implies \quad \Delta S = S(T, V) - S(T_0, V_0)$$

$$= C_V \ln \frac{T}{T_0} + n R \ln \frac{V - n b}{V_0 - n b} .$$

3. $U = U(T, V)$

$$dU = \left(\frac{\partial U}{\partial T}\right)_V dT + \left(\frac{\partial U}{\partial V}\right)_T dV ,$$

(see   1.) $\left(\dfrac{\partial U}{\partial V}\right)_T = T \left(\dfrac{\partial p}{\partial T}\right)_V - p \quad (= 0 \text{ for the ideal gas})$

$$\overset{2.}{=} p + a \frac{n^2}{V^2} - p = a \frac{n^2}{V^2}$$

$$\implies \quad dU = C_V dT + a n^2 \frac{dV}{V^2}$$

$$\implies \quad U = U(T, V) = C_V T - a \frac{n^2}{V} + \text{const} .$$

The interaction of the gas-particles thus provides a volume-dependence of the internal energy, which, in the last analysis, is responsible for the temperature change due to the expansion:

$$U(T_2, V_2) - U(T_1, V_1) = C_V (T_2 - T_1) - a n^2 \left(\frac{1}{V_2} - \frac{1}{V_1}\right) \overset{!}{=} 0$$

$$\implies \quad \Delta T = \frac{a n^2}{C_V} \left(\frac{1}{V_2} - \frac{1}{V_1}\right) .$$

4. According to 2. it holds for reversible adiabatic state changes of the van der Waals gas ($\Delta S = 0$):

$$C_V \ln T + n R \ln(V - n b) = \text{const}_1$$

$$\iff \quad \ln\left[T (V - n b)^{\frac{n R}{C_V}}\right] = \text{const}_2$$

$$\implies \quad T (V - n b)^{\frac{n R}{C_V}} = \text{const}_3 .$$

Compare this adiabatic-equation with that of the ideal gas $T V^{\gamma - 1} = \text{const}$ ($\gamma = \frac{C_p}{C_V} = 1 + \frac{n R}{C_V}$).
Insertion into the equation of state:

$$\left(p + a \frac{n^2}{V^2}\right) (V - n b)^{\frac{n R + C_V}{C_V}} = \text{const}_4 .$$

For the ideal gas these adiabatic-equations read: $p\,V^{\gamma} = $ const.—Finally the equation of state still yields:

$$T^{\frac{nR+C_V}{C_V}}\left(p+a\,\frac{n^2}{V^2}\right)^{-\frac{nR}{C_V}} = \text{const}_5 .$$

This is to be compared with $T^{\gamma}\,p^{1-\gamma} = $ const, valid for the ideal gas!

Solution 2.9.12   Thermal equation of state:

$$V = V_0 - \alpha p + \gamma T .$$

C_p, α, γ are known, as material-specific parameters.

$$\left(\frac{\partial p}{\partial T}\right)_V = \frac{\gamma}{\alpha} ; \quad \left(\frac{\partial V}{\partial T}\right)_p = \gamma .$$

Equation (2.65):

$$C_p - C_V = \left[\left(\frac{\partial U}{\partial V}\right)_T + p\right]\left(\frac{\partial V}{\partial T}\right)_p = T\left(\frac{\partial p}{\partial T}\right)_V\left(\frac{\partial V}{\partial T}\right)_p$$

means in our case here:

$$C_p - C_V = \frac{\gamma^2}{\alpha}\,T$$

and therewith for the heat capacity C_V:

$$C_V = C_p - \frac{\gamma^2}{\alpha}\,T .$$

Internal energy:

$$dU = C_V\,dT + \left(T\left(\frac{\partial p}{\partial T}\right)_V - p\right)dV$$

$$= \left(C_p - \frac{\gamma^2}{\alpha}\,T\right)dT + \frac{V - V_0}{\alpha}\,dV .$$

This can easily be integrated:

$$U(T, V) = U_0 + C_p T - \frac{\gamma^2}{2\alpha}T^2 + \frac{(V - V_0)^2}{2\alpha}$$

$$U(T, p) = U_0 + C_p T + \frac{\alpha}{2}p^2 - \gamma Tp .$$

In the last step we merely inserted the thermal equation of state.

Solution 2.9.13

1. Thermal equation of state:

$$p = \frac{nRT}{V - nb} - a\,\frac{n^2}{V^2} \; .$$

It follows therewith:

$$\left(\frac{\partial p}{\partial T}\right)_V = \frac{nR}{V - nb} \; .$$

Furthermore we utilize:

$$\left(\frac{\partial V}{\partial T}\right)_p = -\left[\left(\frac{\partial T}{\partial p}\right)_V \cdot \left(\frac{\partial p}{\partial V}\right)_T\right]^{-1} = -\left(\frac{\partial p}{\partial T}\right)_V \cdot \left[\left(\frac{\partial p}{\partial V}\right)_T\right]^{-1} \; .$$

It holds for the van der Waals gas:

$$\left(\frac{\partial p}{\partial V}\right)_T = -\frac{nRT}{(V - nb)^2} + 2a\,\frac{n^2}{V^3} \; .$$

It follows then:

$$\left(\frac{\partial V}{\partial T}\right)_p = nR \left[\frac{nRT}{V - nb} - 2a\,\frac{n^2}{V^3}(V - nb)\right]^{-1} \; .$$

Difference of the heat capacities:

$$C_p - C_V \overset{(2.65)}{=} T\left(\frac{\partial p}{\partial T}\right)_V \left(\frac{\partial V}{\partial T}\right)_p = \frac{n^2 R^2 T}{nRT - 2a\frac{n^2}{V^3}(V - nb)^2} \; .$$

Correction with respect to the ideal gas (a and b small):

$$C_p - C_V = \frac{nR}{1 - 2a\frac{n^2}{V^3}\frac{(V-nb)^2}{nRT}} \approx nR\left(1 + 2a\,\frac{n^2}{V^3}\frac{(V - nb)^2}{nRT}\right)$$

$$\approx nR\left(1 + 2a\,\frac{n}{VRT}\right) \; .$$

For the ideal gas it is $C_p - C_V = nR$.

2. Thermal equation of state:

$$p = \frac{nRT}{V - nb} - a\,\frac{n^2}{V^2} \; .$$

We apply (2.58), (2.59):

$$dS = \frac{1}{T} C_V \, dT + \left(\frac{\partial p}{\partial T} \right)_V dV \, .$$

Adiabatic-reversible means $dS = 0$. Therefore:

$$\left(\frac{\partial T}{\partial V} \right)_S = - \left(\frac{\partial p}{\partial T} \right)_V \cdot \frac{T}{C_V} = \frac{-nRT}{C_V(V - nb)} \, .$$

Separation of the variables:

$$\frac{dT}{T} = -\frac{nR}{C_V} \frac{dV}{V - nb} \quad \curvearrowright \quad d\ln T = -\frac{nR}{C_V} d\ln(V - nb) \, .$$

Solution:

$$T = T_0 \left(\frac{V - nb}{V_0 - nb} \right)^{-\frac{nR}{C_V}} \, .$$

Solution 2.9.14 *First law of thermodynamics*

$$\delta Q = dU - \delta W = dU - \varphi \, dq \, .$$

Choose $U = U(T, q)$, i.e.

$$dU = \left(\frac{\partial U}{\partial T} \right)_q dT + \left(\frac{\partial U}{\partial q} \right)_T dq \, .$$

Therewith:

$$\delta Q = \left(\frac{\partial U}{\partial T} \right)_q dT + \left\{ \left(\frac{\partial U}{\partial q} \right)_T - \varphi \right\} dq \, .$$

Second law of thermodynamics
reversible process: $\delta Q = T \, dS$, hence:

$$dS = \frac{1}{T} \left(\frac{\partial U}{\partial T} \right)_q dT + \frac{1}{T} \left\{ \left(\frac{\partial U}{\partial q} \right)_T - \varphi \right\} dq \, .$$

dS and dU are total differentials. Maxwell relation:

$$0 = -\frac{1}{T} \left(\frac{\partial \varphi}{\partial T} \right)_q - \frac{1}{T^2} \left\{ \left(\frac{\partial U}{\partial q} \right)_T - \varphi \right\} \, .$$

That means:

$$\left(\frac{\partial U}{\partial q}\right)_T = \varphi - T\left(\frac{\partial \varphi}{\partial T}\right)_q .$$

Heat quantity:

$$\delta Q = \left(\frac{\partial U}{\partial T}\right)_q dT - T\left(\frac{\partial \varphi}{\partial T}\right)_q dq .$$

Isothermal processing:

$$(\delta Q)_T = 0 - T\left(\frac{\partial \varphi}{\partial T}\right)_q dq .$$

It then remains:

$$(\Delta Q)_{is} = \int_{q_a}^{q_e} \delta Q = -T\frac{d\varphi}{dT} \int_{q_a}^{q_e} dq = -T\frac{d\varphi}{dT}(q_e - q_a) .$$

Solution 2.9.15

1. Basic relation of thermodynamics:

$$dS = \frac{1}{T} dU + \frac{p}{T} dV = C_V \frac{dT}{T} + nR\frac{dV}{V}$$
$$\implies \quad S = S_0 + C_V \ln T + nR \ln V .$$

2. We determine by use of 1. the temperature as function of S and V:

$$\ln T = \frac{1}{C_V}(S - S_0) - \frac{nR}{C_V} \ln V$$

$$\implies \quad T = \frac{\exp\left(\frac{1}{C_V}(S - S_0)\right)}{V^{\frac{nR}{C_V}}}$$

$$\implies \quad U = \frac{C_V}{V^{\frac{nR}{C_V}}} \exp\left[\frac{1}{C_V}(S - S_0)\right] + U_0 = U(S, V) .$$

U as function of the variables S and V even for the ideal gas depends on the volume V. Note that this is **not** a contradiction to the Gay-Lussac experiment.

3.

$$\text{free expansion:} \quad U(T_2, V_2) = U(T_1, V_1) ,$$
$$\text{ideal gas:} \quad U = U(T) \quad \Longrightarrow \quad \Delta T = 0$$
$$\Longrightarrow \quad \text{with 1.:} \quad \Delta S = n R \ln \frac{V_2}{V_1} .$$

Solution 2.9.16 Internal energy:

$$U(T, V) = \int_{}^{T} dT' \, C_V(T') + \varphi(V) = \frac{3}{2} N k_B T - N k_B \frac{N}{V} \left(T^2 \frac{df}{dT} \right) + \varphi(V)$$

$$\Longrightarrow \quad \left(\frac{\partial U}{\partial V} \right)_T = \frac{N^2 k_B T^2}{V^2} \frac{df}{dT} + \varphi'(V) .$$

It follows according to (2.59):

$$\left(\frac{\partial U}{\partial V} \right)_T = T \left(\frac{\partial p}{\partial T} \right)_V - p ,$$

$$\left(\frac{\partial p}{\partial T} \right)_V = \frac{N k_B}{V} \left(1 + \frac{N}{V} f(T) \right) + \frac{N^2 k_B T}{V^2} \frac{df}{dT}$$

$$\Longrightarrow \quad \left(\frac{\partial U}{\partial V} \right)_T = \frac{N^2 k_B T^2}{V^2} \frac{df}{dT} .$$

Comparison with the above expression:

$$\varphi'(V) = 0 \quad \Longleftrightarrow \quad \varphi(V) = \text{const} = U_0$$

$$\Longrightarrow \quad U = \frac{3}{2} N k_B T - N k_B \frac{N}{V} \left(T^2 \frac{df}{dT} \right) + U_0 .$$

We are left with the determination of the entropy!

$$(2.58) + (2.59) \quad \Longrightarrow \quad dS = \frac{C_V}{T} dT + \left(\frac{\partial p}{\partial T} \right)_V dV .$$

Hence:

$$\left(\frac{\partial S}{\partial T} \right)_V = \frac{C_V}{T} ; \quad \left(\frac{\partial S}{\partial V} \right)_T = \left(\frac{\partial p}{\partial T} \right)_V$$

$$\Longrightarrow \quad S(T, V) = \int_{}^{T} dT' \frac{C_V(T')}{T'} + \psi(V) ,$$

$$\frac{C_V}{T} = \frac{3}{2}\frac{Nk_B}{T} - Nk_B\frac{N}{V}\frac{1}{T}\frac{d}{dT}\left(T^2\frac{df}{dT}\right)$$

$$= \frac{3}{2}\frac{Nk_B}{T} - Nk_B\frac{N}{V}\frac{d}{dT}\left(f + T\frac{df}{dT}\right)$$

$$\implies \quad S(T,V) = \frac{3}{2}Nk_B\ln T - \frac{N^2 k_B}{V}\left(f(T) + T\frac{df}{dT}\right) + \psi(V)$$

$$\implies \quad \left(\frac{\partial S}{\partial V}\right)_T = \frac{N^2 k_B}{V^2}\left(f(T) + T\frac{df}{dT}\right) + \psi'(V)\,.$$

Otherwise it holds also:

$$\left(\frac{\partial S}{\partial V}\right)_T = \left(\frac{\partial p}{\partial T}\right)_V = \frac{Nk_B}{V}\left(1 + \frac{N}{V}f(T)\right) + \frac{N^2 k_B T}{V^2}\frac{df}{dT}\,.$$

The comparison yields:

$$\psi'(V) = \frac{Nk_B}{V} \quad \implies \quad \psi(V) = Nk_B\ln V + S_0\,.$$

This means eventually:

$$S(T,V) = \frac{3}{2}Nk_B\ln T + Nk_B\ln V - \frac{N^2 k_B}{V}\left(f(T) + T\frac{df}{dT}\right) + S_0\,.$$

Solution 2.9.17

1. $p = \text{const} = p_0$

$$\Delta W = -\int_{V_1}^{V_2} p\,dV = -p_0\,(V_2 - V_1)\,,$$

$$\delta Q = C_V\,dT + p\,dV\,;\quad dT = \frac{p_0}{nR}\,dV$$

$$\implies \quad \delta Q = \left(\frac{C_V}{nR} + 1\right)p_0\,dV = \frac{\gamma}{\gamma - 1}p_0\,dV$$

$$\implies \quad \Delta Q = \frac{\gamma}{\gamma - 1}p_0\,(V_2 - V_1)\,.$$

Reversible state change:

$$dS = C_V \frac{dT}{T} + \frac{p}{T} dV \; ; \quad \frac{dT}{T} = \frac{dV}{V}$$

$$\implies \quad dS = (C_V + nR) \frac{dV}{V} = C_p \frac{dV}{V}$$

$$\implies \quad \Delta S = C_p \ln \frac{V_2}{V_1} \; .$$

2. $T = \mathrm{const} = T_0$

$$\Delta W = -nRT_0 \int\limits_{V_1}^{V_2} \frac{dV}{V} = -nRT_0 \ln \frac{V_2}{V_1}$$

$$\Delta U = 0 \; , \quad \text{since isotherm} \quad \implies \quad \Delta Q = -\Delta W \; ,$$

$$\text{reversible with } T = \mathrm{const} \quad \implies \quad \Delta S = \frac{1}{T_0} \Delta Q = nR \ln \frac{V_2}{V_1} \; .$$

3. Adiabatic:

$$\text{in addition reversible} \quad \implies \quad \Delta S = \Delta Q = 0 \; ,$$

$$\Delta W = -\int\limits_{V_1}^{V_2} p\, dV = -C \int\limits_{V_1}^{V_2} \frac{dV}{V^\gamma} = \frac{1}{\gamma - 1} \left(\frac{C}{V_2^{\gamma-1}} - \frac{C}{V_1^{\gamma-1}} \right)$$

$$= \frac{1}{\gamma - 1} (p_2 V_2 - p_1 V_1) \; ; \quad p_2 = p_1 \left(\frac{V_1}{V_2} \right)^\gamma$$

$$\implies \quad \Delta W = \frac{p_1}{\gamma - 1} \left[V_2 \left(\frac{V_1}{V_2} \right)^\gamma - V_1 \right] \; .$$

Solution 2.9.18

$$\Delta W = 0 \; , \quad \text{because} \quad \Delta V = 0 \; ,$$

$$\delta Q = C_V dT + p\, dV = C_V dT = \frac{C_V}{nR} V_0 dp$$

$$\implies \quad \Delta Q = \frac{1}{\gamma - 1} V_0 (p_2 - p_1) \; ,$$

$$dS = \frac{C_V}{T} dT \quad \implies \quad \Delta S = \int\limits_{p_1}^{p_2} C_V \frac{dp}{p} = C_V \ln \frac{p_2}{p_1} \; .$$

Solution 2.9.19

1. The general equation (2.59) is valid:

$$\left(\frac{\partial U}{\partial V}\right)_T = T\left(\frac{\partial p}{\partial T}\right)_V - p \,,$$

$$U(T, V) = V\,\varepsilon(T)$$

$$\implies \quad \varepsilon(T) = \alpha\left(T\frac{d\varepsilon(T)}{dT} - \varepsilon(T)\right)$$

$$\Longleftrightarrow \quad (1 + \alpha)\varepsilon(T) = \alpha\,T\frac{d\varepsilon}{dT}$$

$$\Longleftrightarrow \quad \frac{1+\alpha}{\alpha}\frac{dT}{T} = \frac{d\varepsilon}{\varepsilon} \quad \implies \quad \ln T^{\frac{(1+\alpha)}{\alpha}} = \ln\varepsilon + C_0$$

$$\implies \quad U(T, V) = A\,V\,T^{\frac{(1+\alpha)}{\alpha}} \,.$$

2.

$$dS = \frac{1}{T}(dU + p\,dV) = \frac{1}{T}\left(\frac{\partial U}{\partial T}\right)_V dT + \frac{1}{T}\left[p + \left(\frac{\partial U}{\partial V}\right)_T\right]dV$$

$$\implies \quad \left(\frac{\partial S}{\partial T}\right)_V = \frac{1}{T}\left(\frac{\partial U}{\partial T}\right)_V = \frac{1+\alpha}{\alpha}A\,V\,T^{(\frac{1}{\alpha})-1}$$

$$\implies \quad S(T, V) = (1+\alpha)\,A\,V\,T^{\frac{1}{\alpha}} + f(V) \,,$$

$$\left(\frac{\partial S}{\partial V}\right)_T = \frac{1}{T}\left[p + \left(\frac{\partial U}{\partial V}\right)_T\right] = \left(\alpha\,\varepsilon(T) + A\,T^{\frac{(1+\alpha)}{\alpha}}\right)\frac{1}{T} = (\alpha+1)\,A\,T^{\frac{1}{\alpha}}$$

$$\stackrel{!}{=} (1+\alpha)\,A\,T^{\frac{1}{\alpha}} + f'(V) \quad \implies f(V) = \text{const}$$

$$\implies \quad S(T, V) = S_0 + (1+\alpha)\,A\,V\,T^{\frac{1}{\alpha}} \,.$$

Solution 2.9.20

1. Mixing temperature T_m:

The pressure does not change when the separating wall is removed. Thus we have as equations of state:

$$\text{before: } p\,V_{1,2} = n_{1,2}\,R\,T_{1,2} \,,$$

$$\text{after: } p\,V = n\,R\,T_m$$

$$V = V_1 + V_2 \,, \quad n = n_1 + n_2$$

$$\implies \quad nRT_{\mathrm{m}} = R(n_1 T_1 + n_2 T_2)\,,$$

$$T_{\mathrm{m}} = \frac{n_1}{n} T_1 + \frac{n_2}{n} T_2 = \frac{n_1 T_1 + n_2 T_2}{n_1 + n_2}\,.$$

So T_{m} is known.

2. The intermixing is irreversible. The entropy has therefore to be determined via a **reversible auxiliary process**.

a) Isothermal expansion (reversible, see section 2.7) of each of the partial gases:

$$V_{1,2} \quad \longrightarrow \quad V\,, \quad \Delta U_{1,2} = 0\,, \quad \text{since isotherm,}$$

$$\Delta W_{1,2} = -n_{1,2} R T_{1,2} \int_{V_{1,2}}^{V} \frac{dV}{V} = -n_{1,2} R T_{1,2} \ln \frac{V}{V_{1,2}}\,,$$

$$\Delta S_{1,2}^{(a)} = \frac{-\Delta W_{1,2}}{T_{1,2}} = n_{1,2} R \ln \frac{V}{V_{1,2}}\,.$$

The pressure of the two partial gases will have changed.

b) Isochoric temperature change (reversible):

$$T_{1,2} \quad \longrightarrow \quad T_{\mathrm{m}}\,,$$

$$\Delta U = \Delta Q\,, \quad \text{since isochoric,}$$

$$\Delta S_{1,2}^{(b)} = n_{1,2} C_V \int_{T_{1,2}}^{T_{\mathrm{m}}} \frac{dT}{T} = n_{1,2} C_V \ln \frac{T_{\mathrm{m}}}{T_{1,2}}\,.$$

c) Net balance: Entropy of mixing

$$\Delta S = \Delta S_1^{(a)} + \Delta S_2^{(a)} + \Delta S_1^{(b)} + \Delta S_2^{(b)}\,,$$

$$\Delta S = R \left[n_1 \ln \frac{V}{V_1} + n_2 \ln \frac{V}{V_2} \right] + C_V \left[n_1 \ln \frac{T_{\mathrm{m}}}{T_1} + n_2 \ln \frac{T_{\mathrm{m}}}{T_2} \right]\,.$$

3. Let us take $T_1 = T_2$. In the case of identical gases the state of the total system **can not have changed** after the intermixing because of equal pressures and equal temperatures in both the chambers. $\Delta S = 0$ is to be expected. Our calculation, however, yields:

$$\Delta S\,(T_1 = T_2 = T_{\mathrm{m}}) = R \left(n_1 \ln \frac{V}{V_1} + n_2 \ln \frac{V}{V_2} \right)\,.$$

That is **paradoxical** because then the entropy of the system could be made arbitrarily small by squeezing in arbitrarily many separation walls. The entropy would then be a function of the prehistory of the system and therefore **not** a state quantity!

Let $\widehat{C}_V$ be the heat capacity per particle. Then it is generally valid for the entropy of the ideal gas:

$$S = N \left(\widehat{C}_V \ln T + k_B \ln V + C \right) .$$

The *constant* C is independent of T and V, must, however, obviously depend on N. But then we can avoid the *paradox* by the definition

$$S = N \left(\widehat{C}_V \ln T + k_B \ln \frac{V}{N} + C' \right) .$$

The entropy of mixing is unambiguous:

$$\Delta S = S(N, V, T) - [S(N_1, V_1, T_1) + S(N_2, V_2, T_2)]$$

Solution 2.9.21

1.

$$\Delta Q_1 = \int_{T_a}^{T_0} C_p \, dT = C_p \, (T_0 - T_a) .$$

2. The Carnot-machine receives δQ_0 from the heat bath, converts from that δW into work, and gives δQ^* back to the steel block at the intermediate temperature T^* (Fig. A.9). Efficiency of the Carnot machine:

$$\eta_C = 1 - \frac{T^*}{T_0} = 1 + \frac{\delta Q^*}{\delta Q_0} \curvearrowright 0 = \frac{\delta Q^*}{T^*} + \frac{\delta Q_0}{T_0}$$

$$\curvearrowright \delta Q^* = \frac{T^*}{T_0} (-\delta Q_0) = -C_p \, dT^* .$$

Fig. A.9

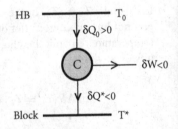

In the last step we have exploited that the process, all in all, shall be performed isobaricly, and that δQ^* is to be counted negative. We can now integrate:

$$\Delta Q_2 = \int_{T_U}^{T_0} \delta Q_0 = -T_0 \int_{T_U}^{T_0} \frac{\delta Q^*}{T^*} = C_p T_0 \int_{T_U}^{T_0} \frac{dT^*}{T^*} \, .$$

That yields:

$$\Delta Q_2 = C_p T_0 \ln \frac{T_0}{T_U} \, .$$

3. (a) **Entropy change for process 1)** At $T = T_0$ the bath gives away heat in a reversible manner:

$$(\Delta S)_{HB}^{(1)} = -\frac{\Delta Q_1}{T_0} = -C_p \left(1 - \frac{T_a}{T_0} \right) \, .$$

The steel block receives **irreversibly** heat to raise its temperature from T_a to T_0. In order to calculate its entropy change we need a *reversible auxiliary process*. This, however, is just the process 2). Therefore

$$(\Delta S)_{block}^{(1)} = \frac{\Delta Q_2}{T_0} = C_p \ln \frac{T_0}{T_U} \, .$$

Total entropy change:

$$\Delta S^{(1)} = C_p \left(-1 + \frac{T_U}{T_0} + \ln \frac{T_0}{T_U} \right) > 0$$

since $\ln x - 1 + 1/x > 0$ for $x > 1$.

Irreversibility creates entropy!

(b) **Entropy change for process 2)**
All sub-steps are now reversible:

$$(\Delta S)_{HB} = \frac{-\Delta Q_2}{T_0} = -C_p \ln \frac{T_0}{T_U}$$

$$(\Delta S)_{Carnot} = 0$$

$$(\Delta S)_{Block} = \int_{T_U}^{T_0} \frac{-\delta Q^*}{T^*} = C_p \ln \frac{T_0}{T_U} \, .$$

That yields for the total entropy change the expected result:

$$\Delta S^{(2)} = 0 .$$

Solution 2.9.22

$$\eta_C = \frac{T_1 - T_2}{T_1} = \frac{1}{6} \quad \Longrightarrow \quad -\Delta W = \eta_C \, \Delta Q_1 = \frac{1}{6} \, kJ$$

Solution 2.9.23

1. Segment $a \to b$:

$$p \sim V \quad \Longrightarrow \quad p = \frac{p_a}{V_a} V \quad \Longrightarrow \quad V_b = \frac{V_a}{p_a} p_b ,$$

ideal gas: $p V = n R T$

$$\Longrightarrow \quad \frac{p_a V_a}{p_b V_b} = \frac{T_a}{T_b} = \frac{p_a^2}{p_b^2} \quad \Longrightarrow \quad T_b = T_a \left(\frac{p_b}{p_a} \right)^2 .$$

Segment $b \to c$:

$$V = \text{const} \quad \Longrightarrow \quad V_c = V_b = \frac{V_a}{p_a} p_b ,$$

$$p_c V_c = n R T_c = p_a V_b \quad \Longrightarrow \quad T_c = p_a V_b \frac{T_b}{p_b V_b}$$

$$\Longrightarrow \quad T_c = T_a \frac{p_b}{p_a} .$$

2. Work done:

$$\Delta W_{ab} = - \int_a^b p \, dV = -\frac{p_a}{V_a} \int_a^b V \, dV = -\frac{p_a}{2 V_a} \left(V_b^2 - V_a^2 \right)$$

$$\Longrightarrow \quad \Delta W_{ab} = -\frac{1}{2} p_a V_a \left[\left(\frac{p_b}{p_a} \right)^2 - 1 \right] < 0 ,$$

$$\Delta W_{bc} = 0 , \quad \text{da } dV = 0,$$

$$\Delta W_{ca} = -p_a (V_a - V_c) = -p_a V_a \left(1 - \frac{p_b}{p_a} \right) > 0 .$$

Internal energy:

The internal energy of the ideal gas depends only on the temperature. From that it follows:

$$\Delta U_{ab} = C_V(T_b - T_a) = C_V T_a \left[\left(\frac{p_b}{p_a}\right)^2 - 1\right],$$

$$\Delta U_{bc} = C_V(T_c - T_b) = C_V T_a \frac{p_b}{p_a}\left(1 - \frac{p_b}{p_a}\right),$$

$$\Delta U_{ca} = C_V(T_a - T_c) = C_V T_a \left(1 - \frac{p_b}{p_a}\right).$$

Heat quantities:

$$\Delta Q_{ab} = \Delta U_{ab} - \Delta W_{ab} = \left(C_V T_a + \frac{1}{2}p_a V_a\right)\left[\left(\frac{p_b}{p_a}\right)^2 - 1\right],$$

$$\Delta Q_{bc} = \Delta U_{bc} - \Delta W_{bc} = C_V T_a \frac{p_b}{p_a}\left(1 - \frac{p_b}{p_a}\right),$$

$$\Delta Q_{ca} = \Delta U_{ca} - \Delta W_{ca} = (C_V T_a + p_a V_a)\left(1 - \frac{p_b}{p_a}\right),$$

$$\Delta Q_{ab} > 0,$$

$$\Delta Q_{bc} < 0; \quad \Delta Q_{ca} < 0.$$

Entropy changes:

$$\Delta S_{b \to c} = \int_b^c \frac{\delta Q}{T} = C_V \int_b^c \frac{dT}{T} = C_V \ln\frac{T_c}{T_b} = C_V \ln\frac{p_a}{p_b},$$

$$\Delta S_{c \to a} = \int_c^a \frac{\delta Q}{T} = C_p \ln\frac{T_a}{T_c} = C_p \ln\frac{p_a}{p_b},$$

S is state quantity: $\oint dS = 0$

$$\implies \quad \Delta S_{a \to b} = -(\Delta S_{b \to c} + \Delta S_{c \to a}) = -(C_V + C_p)\ln\frac{p_a}{p_b},$$

$$C_p = C_V + nR = C_V + \frac{p_a V_a}{T_a}$$

$$\implies \quad \Delta S_{a \to b} = -\left(2 C_V + \frac{p_a V_a}{T_a}\right)\ln\frac{p_a}{p_b}.$$

3.

$$\eta = \frac{\text{total work done}}{\text{accepted heat quantity}} = \frac{-\Delta W}{\Delta Q_{ab}} \, ,$$

$$\Delta W = -\frac{1}{2} p_a V_a \left[\left(\frac{p_b}{p_a} \right)^2 - 1 + 2 - 2\frac{p_b}{p_a} \right]$$

$$= -\frac{1}{2} p_a V_a \left(\frac{p_b}{p_a} - 1 \right)^2 < 0$$

$$\implies \quad \eta = \frac{\frac{1}{2} p_a V_a \, \frac{\frac{p_b}{p_a} - 1}{\frac{p_b}{p_a} + 1}}{C_V T_a + \frac{1}{2} p_a V_a \frac{p_b}{p_a} + 1} = \frac{p_a V_a}{2 C_V T_a + p_a V_a} \frac{p_b - p_a}{p_b + p_a} \, .$$

Solution 2.9.24

1. Temperatures:
   On the adiabatics: $T p^{(1-\gamma)/\gamma} = \text{const}$

$$\implies \quad T_c p_2^{(1-\gamma)/\gamma} = T_d p_1^{(1-\gamma)/\gamma} \, ; \quad T_b p_2^{(1-\gamma)/\gamma} = T_a p_1^{(1-\gamma)/\gamma}$$

$$\implies \quad \frac{T_a - T_d}{T_b - T_c} = \left(\frac{p_2}{p_1} \right)^{(1-\gamma)/\gamma} \, .$$

2. Heat quantities:

$$\Delta Q_{ab} = \Delta Q_{cd} = 0 \, ,$$
$$\Delta Q_{bc} = C_p (T_c - T_b) > 0 \quad (T_c > T_b \text{ follows from the equation of state!}) \, ,$$
$$\Delta Q_{da} = C_p (T_a - T_d) < 0 \, .$$

3. Efficiency:

$$\eta = \frac{-\Delta W}{\Delta Q_{bc}} = 1 + \frac{\Delta Q_{da}}{\Delta Q_{bc}} = 1 - \frac{T_a - T_d}{T_b - T_c} = 1 - \left(\frac{p_1}{p_2} \right)^{(\gamma-1)/\gamma} \, .$$

Solution 2.9.25

1. $\delta Q = T dS$

$$1 \to 2 : \quad T = \text{const} = T_2 \quad \implies \quad \Delta Q_{12} = T_2 (S_2 - S_1) > 0 \, ,$$
$$2 \to 3 : \quad \text{adiabatic, isentropic} \quad \implies \quad \Delta Q_{23} = 0 \, ,$$
$$3 \to 4 : \quad T = \text{const} = T_1 \quad \implies \quad \Delta Q_{34} = T_1 (S_1 - S_2) < 0 \, ,$$
$$4 \to 1 : \quad \text{adiabatic, isentropic} \quad \implies \quad \Delta Q_{41} = 0 \, .$$

2. $0 = \oint dU$, because (thermodynamic) cycle

$$\implies \quad -\Delta W = \oint \delta Q = (T_2 - T_1)(S_2 - S_1)$$

$$\implies \quad \eta = \frac{-\Delta W}{\Delta Q_{12}} = 1 - \frac{T_1}{T_2} = \eta_c .$$

3. The thermodynamic cycle is actually the Carnot process!

Solution 2.9.26

Path (A):

T is a linear function of S:

$$T(S) = -\frac{T_2 - T_1}{S_2 - S_1} S + b .$$

With $T(S_2) = T_1$ it follows:

$$b = T_1 + \frac{T_2 - T_1}{S_2 - S_1} S_2 = \frac{T_2 S_2 - T_1 S_1}{S_2 - S_1} .$$

Hence it holds on the path (A):

$$T(S) = -\frac{T_2 - T_1}{S_2 - S_1} S + \frac{T_2 S_2 - T_1 S_1}{S_2 - S_1} .$$

This yields the heat-exchange contribution:

$$\Delta Q_A = \int_{(A)} dS\, T(S) = -\frac{1}{2}\frac{T_2 - T_1}{S_2 - S_1}\left(S_2^2 - S_1^2\right) + \frac{T_2 S_2 - T_1 S_1}{S_2 - S_1}(S_2 - S_1)$$

$$\curvearrowright \Delta Q_A = \frac{1}{2}(T_2 + T_1)(S_2 - S_1) > 0 .$$

On (A) the system therefore absorbs heat!

Work done (first law of thermodynamics):

$$\Delta W_A = \Delta U_A - \Delta Q_A = U(T_1) - U(T_2) - \Delta Q_A .$$

Here we have exploited that the internal energy of the ideal gas depends only on the temperature.

Path (B)

The process runs isothermally:

$$\Delta U_B = 0 \quad \curvearrowright \quad \Delta W_B = -\Delta Q_B = -T_1(S_1 - S_2) \quad \curvearrowright \quad \Delta Q_B < 0 \, .$$

Path (C):

$$\Delta Q_C = 0 \quad \curvearrowright \quad \Delta W_C = \Delta U_C = U(T_2) - U(T_1) \, .$$

Sum of all the work done:

$$\Delta W = \Delta W_A + \Delta W_B + \Delta W_C$$

$$= U(T_1) - U(T_2) - \frac{1}{2}(T_2 + T_1)(S_2 - S_1) - T_1(S_1 - S_2) + U(T_2) - U(T_1)$$

$$\curvearrowright \quad \Delta W = -\frac{1}{2}(T_2 - T_1)(S_2 - S_1) \, .$$

Efficiency

$$\eta = \frac{-\Delta W}{\Delta Q_A} = \frac{T_2 - T_1}{T_2 + T_1} \, .$$

Solution 2.9.27

Segment $1 \to 2$:

Adiabatic compression of the gas $\Longrightarrow$ temperature rise (via the ignition temperature of the fuel mixture!):

$$\Delta W_{12} = -\int_{V_1}^{V_2} p \, dV = -\int_{V_1}^{V_2} \frac{C_1}{V^\gamma} \, dV = \frac{C_1}{\gamma - 1}\left(V_2^{1-\gamma} - V_1^{1-\gamma}\right)$$

$$= \frac{1}{\gamma - 1}(p_2 V_2 - p_1 V_1) \, ,$$

$$\Delta Q_{12} = 0 \, .$$

Segment $2 \to 3$:

Injection of the fuel (isobaric!):

$$\Delta W_{23} = -p_2 (V_3 - V_2) \, ,$$

$$\Delta Q_{23} = C_p (T_3 - T_2) \, ,$$

$$p_2 V_2 = n R T_2 \, ,$$

$$p_2 V_3 = n R T_3$$

$$\implies \quad T_3 - T_2 = \frac{p_2}{n R} (V_3 - V_2) \,,$$

$$\Delta Q_{23} = \frac{C_p}{n R} p_2 (V_3 - V_2)$$

$$= \frac{\gamma}{\gamma - 1} p_2 (V_3 - V_2) > 0 \,.$$

Segment $3 \to 4$:

Expansion along an adiabatic curve (work done):

$$\Delta W_{34} = \frac{C_2}{\gamma - 1} \left(V_4^{1-\gamma} - V_3^{1-\gamma} \right) = \frac{1}{\gamma - 1} (p_4 V_1 - p_2 V_3) \,,$$

$$\Delta Q_{34} = 0 \,.$$

Segment $4 \to 1$:

Ejection of the residual gas:

$$\Delta W_{41} = 0 \,,$$

$$\Delta Q_{41} = C_V (T_1 - T_4) \,,$$

$$p_4 V_1 = n R T_4 \,,$$

$$p_1 V_1 = n R T_1$$

$$\implies \quad T_1 - T_4 = \frac{V_1}{n R} (p_1 - p_4) \,,$$

$$\Delta Q_{41} = \frac{C_V}{n R} V_1 (p_1 - p_4)$$

$$= \frac{1}{\gamma - 1} V_1 (p_1 - p_4) < 0 \,.$$

Net balance:

$$\Delta W = \oint \delta W = \frac{1}{\gamma - 1} (p_2 V_2 - p_1 V_1 + p_4 V_1 - p_2 V_3) - p_2 (V_3 - V_2)$$

$$= p_2 (V_3 - V_2) \frac{\gamma}{1 - \gamma} + \frac{V_1}{\gamma - 1} (p_4 - p_1)$$

$$= \frac{1}{\gamma - 1} [V_1 (p_4 - p_1) - \gamma p_2 (V_3 - V_2)] \,.$$

Solution 2.9.28

1. $0 \to 1$:   aspiration, piston is shifted, volume-expansion at constant pressure,
   $1 \to 2$:   densification, adiabatic compression, thereby pressure rise,
   $2 \to 3$:   ignition, pressure rise at constant volume,
   $3 \to 4$:   adiabatic expansion, thereby work done,
   $4 \to 1$:   opening of the outlet valves, pressure decrease at constant volume,
   $1 \to 0$:   expulsion of the residual gas.
   ($0 \to 1$, $1 \to 0$ do not belong to the thermodynamic cycle!)

2. $\Delta U = \Delta Q + \Delta W = 0$, since cycle

$$\implies \quad \Delta W = -\Delta Q \, ,$$
$$\Delta Q = C_V \, (T_3 - T_2) + C_V \, (T_1 - T_4) = \Delta Q_{23} + \Delta Q_{41} \, .$$

Equation of state:

$$p_2 \, V_2 = n \, R \, T_2 \, ; \quad p_3 \, V_2 = n \, R \, T_3$$
$$\implies \quad T_3 > T_2 \, , \quad \text{because} \quad p_3 > p_2 \, ,$$
$$p_1 \, V_1 = n \, R \, T_1 \, ; \quad p_4 \, V_1 = n \, R \, T_4$$
$$\implies \quad T_4 > T_1 \, , \quad \text{because} \quad p_4 > p_1$$
$$\implies \quad \Delta Q_{23} = C_V \, (T_3 - T_2) > 0 \, ,$$
$$\Delta Q_{41} = C_V \, (T_1 - T_4) < 0 \, .$$

Adiabatic equations:

$$T_1 \, V_1^{\gamma - 1} = T_2 \, V_2^{\gamma - 1} \quad \implies \quad T_1 < T_2 \, ,$$
$$T_3 \, V_2^{\gamma - 1} = T_4 \, V_1^{\gamma - 1} \quad \implies \quad T_4 < T_3 \, .$$

T_3 is thus the highest, T_1 the lowest temperature of the cycle!

$$(T_1 - T_4) \, V_1^{\gamma - 1} = (T_2 - T_3) \, V_2^{\gamma - 1}$$
$$\implies \quad \Delta Q = -\Delta W = C_V \, (T_3 - T_2) \left[1 - \left(\frac{V_2}{V_1} \right)^{\gamma - 1} \right] \, .$$

3. On the segment $2 \to 3$ the system absorbs heat, therefore:

$$\eta = \frac{-\Delta W}{\Delta Q_{23}} = 1 - \left(\frac{V_2}{V_1} \right)^{\gamma - 1} = 1 - \frac{T_1}{T_2} \, .$$

4. The Carnot machine between the heat baths $HB(T_1)$ and $HB(T_3)$ has the efficiency

$$\eta_C = 1 - \frac{T_1}{T_3}$$

$$\implies \quad \eta_{otto} < \eta_C .$$

$T_2 \rightarrow T_3$ is not possible with the *Otto motor*, because then the *ignition* would be left out.

Solution 2.9.29

$\underline{1 \rightarrow 2}$: Heat $Q_D = Q_D^{(1)} + Q_D^{(2)}$ is taken from the heat bath $HB(T)$ and applied to overcome the cohesive forces $(Q_D^{(1)})$ and for the expansion $(Q_D^{(2)})$:

$$\Delta W_{12} = -(p + \Delta p)(V_2 - V_1) .$$

$\underline{2 \rightarrow 3}$: Hardly any work is done because of negligible volume-change:

$$\Delta W_{23} \approx 0 .$$

$\underline{3 \rightarrow 4}$: Isothermal condensation:

$$\Delta W_{34} = -p(V_1 - V_2) .$$

$Q_D^{(1)}$ goes via condensation into the heat bath $HB(T - \Delta T)$.
$\underline{4 \rightarrow 1}$: $\Delta W_{41} \approx 0$, because of an only unimportant volume-change.

$$\implies \quad \Delta W = \oint \delta W = -\Delta p (V_2 - V_1) < 0 .$$

Efficiency:

$$(Carnot) \quad \eta = 1 - \frac{T - \Delta T}{T} = \frac{\Delta T}{T} \overset{!}{=} \frac{-\Delta W}{Q_D} = \frac{\Delta p (V_2 - V_1)}{Q_D}$$

$$\implies \quad \frac{\Delta p}{\Delta T} = \frac{Q_D}{T(V_2 - V_1)}$$

Solution 2.9.30 According to (2.73) it holds for adiabatic state changes:

$$(dT)_{ad} = -T \frac{\beta(T)}{C_V \kappa_T} (dV)_{ad} .$$

Because of $\beta(T = 4\,^\circ\text{C}) = 0$ the **adiabatic** cooling down of water from 6 to $4\,^\circ\text{C}$ is not possible. The described Carnot process is actually not realizable. Therefore it can not be about a contradiction.

Solution 2.9.31 Works done:

$$\Delta W_{12} = -\int_{V_1}^{V_2} p(V)\,dV = -nRT_2 \ln \frac{V_2}{V_1} < 0 \,,$$

$$\Delta W_{23} = 0 \,, \quad \text{da} \quad dV = 0 \,,$$

$$\Delta W_{34} = -\int_{V_2}^{V_1} p(V)\,dV = -nRT_1 \ln \frac{V_1}{V_2} > 0 \,,$$

$$\Delta W_{41} = 0$$

$$\implies \quad \Delta W = -nR\,(T_2 - T_1) \ln \frac{V_2}{V_1} \,.$$

Equation of state:

$$\begin{aligned} p_4\,V_1 &= nRT_1 \\ p_1\,V_1 &= nRT_2 \end{aligned} \quad \implies \quad T_2 > T_1 \quad \implies \quad \Delta W < 0 \,.$$

Heat quantities:
(during isothermal state changes the internal energy of the ideal gas remains constant!)

$$\Delta Q_{12} = -\Delta W_{12}$$

$$\Delta Q_{23} = C_V\,(T_1 - T_2) < 0 \,,$$

$$\Delta Q_{34} = -\Delta W_{34} \,,$$

$$\Delta Q_{41} = C_V\,(T_2 - T_1) > 0 \,.$$

Efficiency:

$$\eta = \frac{-\Delta W}{\Delta Q} \,.$$

ΔQ is the amount of heat supplied to the system:

$$\Delta Q = \Delta Q_{12} + \Delta Q_{41} \neq \Delta Q_{12} \,,$$

$$\Delta Q = nRT_2 \ln \frac{V_2}{V_1} + C_V\,(T_2 - T_1)$$

$$\implies \quad \eta = \frac{T_2 - T_1}{T_2 + \frac{C_V(T_2 - T_1)}{nR\ln V_2/V_1}} < \eta_C \,.$$

Fig. A.10

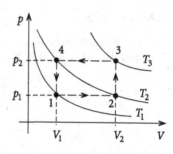

Fig. A.11

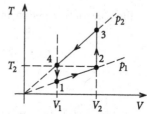

Fig. A.12

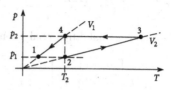

Is that a contradiction to the statement, proven in Sect. 2.5, that η_C is reached by all machines which work reversibly and periodically between two heat baths?

Solution 2.9.32

1. Isotherms: $p \propto 1 \, / \, V$ (equation of state!)

   a) $p_1 V_1 = n R T_1$,
   b) $p_1 V_2 = n R T_2$,
   c) $p_2 V_2 = n R T_3$,
   d) $p_2 V_1 = n R T_4$.

   Because of $p_1 V_2 = p_2 V_1$ we have $T_2 = T_4$. There are thus only three different temperatures T_1, T_2, T_3 (Fig. A.10). We need three isotherms for the representation.

2. Isobares: $T \propto V$ (Fig. A.11)

$$p(1) = p(2) = p_1 \; ; \quad p(3) = p(4) = p_2 \; ,$$
$$V(1) = V(4) = V_1 \; ; \quad V(2) = V(3) = V_2 \; .$$

3. Isochores: $p \propto T$ (Fig. A.12)

Solution 2.9.33

1. An elongation of the thread means work is done on the system which is therefore to be counted as positive. It comes out therewith the following analogy to the gas:

$$Z\,dL \quad \Longleftrightarrow \quad -p\,dV\,,$$
$$Z \quad \Longleftrightarrow \quad -p\,,$$
$$L \quad \Longleftrightarrow \quad V\,.$$

First law of thermodynamics:

$$dU = \delta Q + Z\,dL\,,$$
$$dS = \frac{dU}{T} - \frac{Z}{T}\,dL\,,$$
$$U = U(T,L) \quad \Longrightarrow \quad dS = \frac{1}{T}\left(\frac{\partial U}{\partial T}\right)_L dT + \left[\frac{1}{T}\left(\frac{\partial U}{\partial L}\right)_T - \frac{Z}{T}\right]dL\,.$$

dS is a total differential. From that it follows:

$$\frac{1}{T}\left[\frac{\partial}{\partial L}\left(\frac{\partial U}{\partial T}\right)_L\right]_T \overset{!}{=} \left\{\frac{\partial}{\partial T}\left[\frac{1}{T}\left(\frac{\partial U}{\partial L}\right)_T - \frac{Z}{T}\right]\right\}_L \qquad \text{(integrability conditions)}$$

$$\Longrightarrow \quad \frac{1}{T^2}\left(\frac{\partial U}{\partial L}\right)_T = -\left[\frac{\partial}{\partial T}\left(\frac{Z}{T}\right)\right]_L\,,$$

$$\left[\frac{\partial}{\partial T}\left(\frac{Z}{T}\right)\right]_L = \left(\frac{\partial}{\partial T}\frac{L-L_0}{\alpha}\right)_L = 0$$

$$\Longrightarrow \quad \left(\frac{\partial U}{\partial L}\right)_T = 0 \quad \Longleftrightarrow \quad U(T,L) \equiv U(T)\,.$$

Heat capacity:

$$C_L(T,L) = \left(\frac{\delta Q}{dT}\right)_L = \left(\frac{\partial U}{\partial T}\right)_L \equiv C_L(T)\,.$$

According to the scope of the exercise we have especially for $L = L_0$:

$$C_{L_0}(T) = C_{L_0} = C > 0\,.$$

Because of $C_L(T) = C_{L_0}(T)$ it follows then for arbitrary L the assertion:

$$C_L(T) \equiv C > 0\,.$$

2. From part 1. it follows already:

$$U(T) = CT + U_0 \,,$$

$$dS = \frac{dU}{T} - \frac{Z}{T}\,dL = C\,\frac{dT}{T} - \frac{L - L_0}{\alpha}\,dL$$

$\Longrightarrow$ entropy:

$$S(T, L) = C \ln T - \frac{1}{2\alpha}\,(L - L_0)^2 + S_0 \,.$$

Adiabatic equations:

$$dS = 0 \quad\Longrightarrow\quad C\,\frac{dT}{T} = \frac{L - L_0}{\alpha}\,dL$$

$$\Longleftrightarrow \quad d \ln T = \frac{L - L_0}{\alpha\,C}\,dL$$

$$\Longleftrightarrow \quad T(L) = D \exp\left[\frac{(L - L_0)^2}{2\,\alpha\,C}\right]\,,$$

The constant D takes different values on different adiabatic curves!

$$D = T(L_0) = \exp\left[(S - S_0)\,/\,C\right] = D(S)\,.$$

$$Z(L) = \frac{1}{\alpha}(L - L_0)\,T(L) = \frac{D}{\alpha}\,(L - L_0)\exp\left[\frac{(L - L_0)^2}{2\,\alpha\,C}\right]\,.$$

3. See Fig. A.13

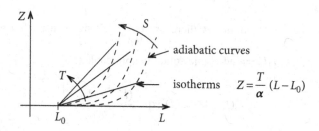

Fig. A.13

4.

$$C_Z = T \left(\frac{\partial S}{\partial T} \right)_Z .$$

Thus we need $S = S(T, Z)$:

$$dS = \frac{C}{T} dT - \frac{Z}{T} dL ,$$

$$L = L(T, Z) = L_0 + \alpha \frac{Z}{T} \implies dL = -\frac{\alpha Z}{T^2} dT + \frac{\alpha}{T} dZ$$

$$\implies dS = \left(\frac{C}{T} + \alpha \frac{Z^2}{T^3} \right) dT - \frac{\alpha Z}{T^2} dZ$$

$$\implies C_Z = C + \alpha \frac{Z^2}{T^2} .$$

5.

$$\Delta L = \alpha Z \left(\frac{1}{T_2} - \frac{1}{T_1} \right) < 0 , \quad \text{falls} \quad \alpha > 0 ,$$

$$\beta = \frac{-\Delta W}{\Delta Q} = 1 - \frac{\Delta U}{\Delta Q} < 1 , \quad \begin{array}{l} \text{since during the heating also} \\ \text{the internal energy will change,} \end{array}$$

$$\Delta U = C (T_2 - T_1) ,$$

$$\Delta Q = \int_{T_1}^{T_2} C_Z \, dT = C (T_2 - T_1) - \alpha Z^2 \left(\frac{1}{T_2} - \frac{1}{T_1} \right)$$

$$\implies \beta = \frac{1}{1 + \frac{C}{\alpha Z^2} T_1 T_2} < 1 .$$

The supplied heat quantity is therefore **not exclusively** used for mechanical work!

6. *Thermally isolated* $\implies$ adiabatic equation applicable (see 2.).

$$T (L_{1,2}) = D \exp \left[\frac{(L_{1,2} - L_0)^2}{2 \alpha C} \right] ,$$

$$L_2 > L_1 \implies T (L_2) > T (L_1) .$$

The thread therefore heats up.

Fig. A.14

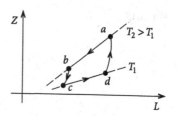

Solution 2.9.34

1. The Carnot process consists of two isotherms and two adiabatic curves (Fig. A.14). It works as *heat engine* if it extracts heat from a heat reservoir and gives away a part of it for doing work:

$$\oint \delta W \overset{!}{<} 0 \; ; \quad \oint \delta Q > 0 \; .$$

Because of $\delta W = Z \, dL$ the direction of the path is as sketched in Fig. A.14.

2. First law of thermodynamics:

$$\delta Q = dU - Z \, dL \; ; \quad dU = C \, dT \; ,$$

$$\Delta Q_2 = \int_a^b \delta Q = -\int_a^b Z(T, L) \, dL = \quad \text{(isotherm!)}$$

$$= -\frac{T_2}{\alpha} \int_{L_a}^{L_b} (L - L_0) \, dL = -\frac{T_2}{2\alpha} \left[(L_b - L_0)^2 - (L_a - L_0)^2 \right] \; ,$$

$$L_a > L_b > L_0 \quad \Longrightarrow \quad \Delta Q_2 > 0 \; .$$

The system absorbs heat on the segment $a \to b$. Analogously one finds:

$$\Delta Q_1 = -\frac{T_1}{2\alpha} \left[(L_d - L_0)^2 - (L_c - L_0)^2 \right]$$

$$L_0 < L_c < L_d \quad \Longrightarrow \quad \Delta Q_1 < 0 \; .$$

On the segment $c \to d$ the system transfers heat into the heat bath $HB(T_1)$. Using part 2. of exercise 2.9.33 one gets:

$$(L_{a,b,c,d} - L_0)^2 = 2\alpha \left[C \ln T_{a,b,c,d} - (S_{a,b,c,d} - S_0) \right] \; .$$

Fig. A.15

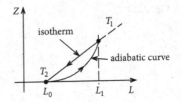

On the adiabatic curves we have:

$$S_a = S_d \quad ; \quad S_b = S_c$$

$$\implies \quad (L_b - L_0)^2 - (L_a - L_0)^2 = 2\alpha \left[\dot{S}_a - S_b \right] ,$$

$$(L_d - L_0)^2 - (L_c - L_0)^2 = 2\alpha \left[S_c - S_d \right] = 2\alpha \left[S_b - S_a \right] .$$

Efficiency:

$$\eta = \frac{\Delta Q_1 + \Delta Q_2}{\Delta Q_2} = 1 - \frac{T_1}{T_2} = \eta_C .$$

3. The point at L_0 is ambiguous on the isotherm as well as on the adiabatic curve
   (Fig. A.15). Away from this point it is on the isotherm $T = T_1$. For the adiabatic
   curve we have according to part 2. of exercise 2.9.33:

$$T = D(S) \exp\left[\frac{(L - L_0)^2}{2\alpha C} \right] = T(L)$$

$$\implies \quad T_2 = (T(L_0))_{ad} = D ,$$

$$T_1 = (T(L_1))_{ad} = D \exp\left[\frac{(L_1 - L_0)^2}{2\alpha C} \right]$$

$$= T_2 \exp\left[\frac{(L_1 - L_0)^2}{2\alpha C} \right] > T_2 .$$

At $L = L_0$ the system must be cooled down from T_1 to T_2. This process is
irreversible if one uses the heat bath $HB(T_2)$ for that. We therefore expect

$$\eta < \eta_C = 1 - \frac{T_2}{T_1} .$$

Work done:

$$\Delta W = \int_{is} Z\,dL + \int_{ad} Z\,dL$$

$$= \frac{T_1}{\alpha} \int_{L_1}^{L_0} (L - L_0)\,dL + \frac{D}{\alpha} \int_{L_0}^{L_1} (L - L_0)\, e^{\frac{(L-L_0)^2}{2\alpha C}}\,dL$$

$$= -\frac{T_1}{2\alpha}(L_1 - L_0)^2 + \frac{T_2}{\alpha}\alpha C \int_{L_0}^{L_1} dL \frac{d}{dL} e^{\frac{(L-L_0)^2}{2\alpha C}} \,,$$

$$\Delta W = -\frac{T_1}{2\alpha}(L_1 - L_0)^2 + C T_2 \left\{ \exp\left[\frac{(L_1 - L_0)^2}{2\alpha C}\right] - 1 \right\} \,.$$

Amount of heat:

$$\Delta Q = (\Delta Q)_{\text{is}} = \int_{\text{is}} (dU - Z\,dL) = -\frac{T_1}{\alpha} \int_{L_1}^{L_0} (L - L_0)\,dL$$

$$\implies \quad \Delta Q = \frac{T_1}{2\alpha}(L_1 - L_0)^2 > 0 \,, \quad \text{is taken by the system!}$$

Efficiency:

$$\eta = \frac{-\Delta W}{\Delta Q} = 1 - \frac{T_2}{T_1} \frac{e^{\frac{(L_1-L_0)^2}{2\alpha C}} - 1}{\frac{(L_1-L_0)^2}{2\alpha C}} \,,$$

$$\frac{1}{x}(e^x - 1) = 1 + \frac{1}{2!}x > 1 \,, \quad (x > 0) \,.$$

Thus it is as expected:

$$\eta < \eta_C = 1 - \frac{T_2}{T_1} \,.$$

Solution 2.9.35

$\Delta W = 0$, since the change of the length takes place against $Z = 0$,
$\Delta Q = 0$, since no heat exchange with the surroundings occurs.

A reversible auxiliary process would yield the following entropy change:
Part 2. of exercise 2.9.33 had given:

$$S(T, L) = C \ln T - \frac{1}{2\alpha}(L - L_0)^2 + S_0$$

$$\implies \quad \Delta S = S(T, L_0) - S(T, L) = \frac{1}{2\alpha}(L - L_0)^2 = \frac{1}{2}\alpha\frac{Z^2}{T^2} \,.$$

Reversible auxiliary process:

Isothermal state change by contacting a heat bath $HB(T)$, quasi-static procedure. Thereby $Z = 0$, i.e. $L \to L_0$:

$$dU = 0 \quad \Longrightarrow \quad \delta Q = -\delta W = -Z\,dL = T\,dS$$

$$\Longrightarrow \quad \Delta S = -\frac{1}{T}\int_L^{L_0} Z\,dL' = -\frac{1}{\alpha}\int_L^{L_0}(L' - L_0)\,dL'$$

$$= \frac{(L - L_0)^2}{2\alpha} = \frac{1}{2}\alpha\,\frac{Z^2}{T^2}$$

That was to be shown!

Solution 2.9.36

1.

$$\delta Q^{(m)} = dU^{(m)} - \delta W^{(m)}\ ,$$

$$\delta W^{(m)} = \mu_0\,V\,H\,dM$$

$$\Longrightarrow \quad C_M^{(m)} = \left(\frac{\delta Q^{(m)}}{dT}\right)_M = \left(\frac{\partial U^{(m)}}{\partial T}\right)_M ,$$

$$C_H^{(m)} = C_M^{(m)} + \left[\left(\frac{\partial U^{(m)}}{\partial M}\right)_T - \mu_0 H V\right]\left(\frac{\partial M}{\partial T}\right)_H .$$

It holds for the entropy of the paramagnetic moment system:

$$dS^{(m)} = \frac{1}{T}\left(\frac{\partial U^{(m)}}{\partial T}\right)_M dT + \frac{1}{T}\left[\left(\frac{\partial U^{(m)}}{\partial M}\right)_T - \mu_0 VH\right]dM\ .$$

Integrability condition for the total differential $dS^{(m)}$:

$$\frac{1}{T}\left(\frac{\partial^2 U^{(m)}}{\partial M\,\partial T}\right) \stackrel{!}{=} -\frac{1}{T^2}\left[\left(\frac{\partial U^{(m)}}{\partial M}\right)_T - \mu_0 VH\right]$$

$$+ \frac{1}{T}\left[\left(\frac{\partial^2 U^{(m)}}{\partial T\,\partial M}\right) - \mu_0 V\left(\frac{\partial H}{\partial T}\right)_M\right].$$

We utilize now the integrability condition for $dU^{(m)}$ and insert the equation of state (Curie law):

$$\left(\frac{\partial U^{(m)}}{\partial M}\right)_T = \mu_0 VH - T\mu_0 V\left(\frac{\partial H}{\partial T}\right)_M = 0\ .$$

This means

$$dU^{(m)} = \left(\frac{\partial U^{(m)}}{\partial T}\right)_M dT = C_M^{(m)} dT$$

or for the heat quantity:

$$\delta Q^{(m)} = C_M^{(m)} dT - \mu_0 VH \left[\left(\frac{\partial M}{\partial T}\right)_H dT + \left(\frac{\partial M}{\partial H}\right)_T dH\right]$$

$$= C_H^{(m)} dT - \mu_0 VH \left(\frac{\partial M}{\partial H}\right)_T dH \ .$$

That shows the assertion.

2a.

$$\delta Q^{(m)} = 0$$

$$\Longrightarrow \quad \frac{dT}{dH} = \mu_0 V H \frac{\left(\frac{\partial M}{\partial H}\right)_T}{C_H^{(m)}} = \frac{\widehat{C}\frac{1}{T} \cdot H}{\widehat{C}\frac{H^2 + H_r^2}{T^2}}$$

$$\Longrightarrow \quad \frac{dT}{dH} = \frac{TH}{H^2 + H_r^2} \ .$$

This can easily be integrated:

$$d\ln T = \frac{dT}{T} = \frac{H\, dH}{H^2 + H_r^2} = \frac{1}{2}\frac{d}{dH}\ln\left(H^2 + H_r^2\right) dH \ .$$

It follows with $T_0 = T(H = 0)$:

$$\ln\frac{T}{T_0} = \frac{1}{2}\ln\frac{H^2 + H_r^2}{H_r^2}$$

$$\Longrightarrow \quad T(H) = T_0 \sqrt{\frac{H^2 + H_r^2}{H_r^2}} \ .$$

2b. Thermal equilibrium means:

$$\delta Q^{(m)} = -\delta Q_L = -C_L\, dT \ ,$$

C_L is thought to be known,

$$\implies \quad -\left(C_L + C_H^{(m)}\right) dT = -\mu_0 \, V \, H \left(\frac{\partial M}{\partial H}\right)_T dH$$

$$\implies \quad \frac{dT}{dH} = \mu_0 \, V \, \frac{\widehat{C} \frac{H}{T}}{C_L + C_H^{(m)}} \approx \mu_0 \, V \, \frac{\widehat{C}}{C_L} \frac{H}{T} \ .$$

since $C_L \gg C_H^{(m)}$

This is again easily integrated:

$$\frac{1}{2}\left(T^2 - T_0^2\right) = \mu_0 \, V \, \frac{\widehat{C}}{C_L} \frac{1}{2} H^2$$

$$\implies \quad T(H) = \sqrt{\mu_0 \, V \, \frac{\widehat{C}}{C_L} H^2 + T_0^2} \ .$$

3a. That is the situation of 2a.:

$$T_0 = T^* \sqrt{\frac{H_r^2}{H^{*2} + H_r^2}} < T^* \quad \text{(cooling down!)}.$$

3b.

$$\Longleftrightarrow \quad -\int_{T_0}^{T_f} \delta Q^{(m)} \overset{!}{=} \int_{T^*}^{T_f} \delta Q_L \quad (H = \text{const} = 0)$$

$$\Longleftrightarrow \quad -\int_{T_0}^{T_f} C_H^{(m)}(H = 0)\, dT = \int_{T^*}^{T_f} C_L \, dT$$

$$\Longleftrightarrow \quad -\widehat{C} \, \mu_0 \, V \, H_r^2 \int_{T_0}^{T_f} \frac{dT}{T^2} = C_L \left(T_f - T^*\right)$$

$$\Longleftrightarrow \quad \widehat{C} \, \mu_0 \, V \, H_r^2 \left(\frac{1}{T_f} - \frac{1}{T_0}\right) = C_L \left(T_f - T^*\right) \ .$$

The temperature of the lattice will change only very little with the above heat exchange because of its large heat capacity. Thus we can replace on the left-hand side approximately T_f by T^*:

$$T_f - T^* = \frac{\widehat{C}\,\mu_0\,V}{C_L}H_r^2\left(\frac{1}{T^*} - \frac{1}{T_0}\right)$$

$$= \frac{\widehat{C}\,\mu_0\,V}{C_L}\frac{H_r^2}{T^*}\left(1 - \frac{\sqrt{H_r^2 + H^{*2}}}{H_r}\right) < 0.$$

4. That is now the situation of 2b., for which it remains to solve because of $C_L \gg C_H^{(m)}$:

$$\frac{dT}{dH} \approx \mu_0\,V\frac{\widehat{C}}{C_L}\frac{H}{T}$$

$$\implies \widehat{T}_f^2 - T^{*2} = -\mu_0\,V\frac{\widehat{C}}{C_L}H^{*2}$$

$$\iff \left(\widehat{T}_f - T^*\right)\left(\widehat{T}_f + T^*\right) = -\mu_0\,V\frac{\widehat{C}}{C_L}H^{*2}.$$

From the same reasons as in 3b. we can take

$$\widehat{T}_f + T^* \approx 2\,T^*,$$

getting therewith:

$$\implies \widehat{T}_f - T^* \approx -\mu_0\,V\frac{\widehat{C}}{2\,C_L}\frac{H_*^2}{T^*} < 0.$$

Even now it gives rise to a cooling!

5a. The process 3. is irreversible, namely because of the heat exchange between the system of moments and the crystal (part 2.). The process 4. is reversible. The total system is thermally isolated:

$$\Delta Q = \Delta Q^{(m)} + \Delta Q_L = 0.$$

Generally it holds in such a case:

$$dS \geq \frac{\delta Q}{T} = 0,$$

'=' for reversible; '>' for irreversible $\Longrightarrow$ process 3.: $\Delta S > 0$; process 4.: $\Delta S = 0$

$$\Longrightarrow \quad S_{\mathrm{f}} > \widehat{S}_{\mathrm{f}} \ .$$

5b. By resolving $S = S(T, H)$ for T,

$$T = T(S, H) \ ,$$

one finds for the final states:

$$T_{\mathrm{f}} = T\left(S_{\mathrm{f}}, 0\right) \neq \widehat{T}_{\mathrm{f}} = T(\widehat{S}_{\mathrm{f}}, 0) \ .$$

We assert that

$$\widehat{T}_{\mathrm{f}} \geq T_{\mathrm{f}}$$

That is indeed so if it holds:

$$0 \geq \mu_0\, V \frac{\widehat{C}}{C_L} \frac{1}{T^*} \left(-\frac{1}{2} H^{*2} - H_{\mathrm{r}}^2 + H_{\mathrm{r}}\sqrt{H_{\mathrm{r}}^2 + H^{*2}} \right)$$

$$\Longleftrightarrow \quad H^{*2} \geq 2\left(H_{\mathrm{r}}\sqrt{H_{\mathrm{r}}^2 + H^{*2}} - H_{\mathrm{r}}^2 \right)$$

$$\Longleftrightarrow \quad H^{*2} + 2\,H_{\mathrm{r}}^2 \geq 2\,H_{\mathrm{r}}\sqrt{H_{\mathrm{r}}^2 + H^{*2}}$$

$$\Longleftrightarrow \quad H^{*4} + 4\,H^{*2}\,H_{\mathrm{r}}^2 + 4\,H_{\mathrm{r}}^4 \geq 4\,H_{\mathrm{r}}^4 + 4\,H_{\mathrm{r}}^2\,H^{*2}$$

$$\Longleftrightarrow \quad H^{*4} \geq 0 \quad \mathrm{q.\,e.\,d.}$$

The equal sign $(\widehat{T}_{\mathrm{f}} = T_{\mathrm{f}})$ is valid only for $H^* = 0$.

Section 3.9

Solution 3.9.1

1. Free energy: $F = F(T, V)$

$$\left(\frac{\partial F}{\partial T} \right)_V = -S(T, V) \ ; \quad \left(\frac{\partial F}{\partial V} \right)_T = -p(T, V) \ .$$

We integrate the first equation:

$$
F(T,V) \;=\; -\int_{T_0}^{T} dT'\, S(T',V) + f(V) = -\frac{R\,V_0}{V}\,\frac{1}{T_0^a}\int_{T_0}^{T} dT'\, T'^a + f(V)
$$

$$
\overset{(a\neq -1)}{=} -\frac{R\,V_0}{V\,T_0^a}\,\frac{1}{a+1}\left(T^{a+1} - T_0^{a+1}\right) + f(V)\;.
$$

Intermediate result:

$$
F(T,V) = -\frac{R\,V_0}{V}\,\frac{T}{a+1}\left(\frac{T}{T_0}\right)^a + \frac{R\,V_0}{V}\,\frac{T_0}{a+1} + f(V)\;.
$$

Because of

$$
dF = -S\,dT - p\,dV = -S\,dT + \delta W
$$

we get for **isothermal** state changes:

$$
(dF)_T = (\delta W)_T\;.
$$

That we utilize for $T = T_0$:

$$
\left(\frac{\partial F}{\partial V}\right)_{T_0} = \left(\frac{\delta W}{\partial V}\right)_{T_0} = \frac{R\,T_0}{V}
$$

$$
= \left\{\frac{R\,V_0}{V^2}\,\frac{T}{a+1}\left(\frac{T}{T_0}\right)^a - \frac{R\,T_0\,V_0}{V^2(a+1)} + f'(V)\right\}_{T=T_0} = f'(V)
$$

$$
\implies\; f(V) = R\,T_0 \ln\frac{V}{V_0} + f(V_0)\;.
$$

This fixes eventually the free energy:

$$
F(T,V) = R\,\frac{V_0}{V}\,\frac{T_0}{a+1}\left[1 - \left(\frac{T}{T_0}\right)^{a+1}\right] + R\,T_0 \ln\frac{V}{V_0} + F(T_0, V_0)\;.
$$

2. Equation of state:

$$
p = -\left(\frac{\partial F}{\partial V}\right)_T = R\,\frac{V_0}{V^2}\,\frac{T_0}{a+1}\left[1 - \left(\frac{T}{T_0}\right)^{a+1}\right] - \frac{R\,T_0}{V}\;.
$$

3. Work done:

$$\Delta W_T = -\int\limits_{V_0}^{V} p(T, V')\, dV'$$

$$= R V_0 \frac{T_0}{a+1}\left[1 - \left(\frac{T}{T_0}\right)^{a+1}\right]\left(\frac{1}{V} - \frac{1}{V_0}\right) + R T_0 \ln \frac{V}{V_0}$$

$$\implies \quad \Delta W_T = \frac{R T_0}{a+1}\left[1 - \left(\frac{T}{T_0}\right)^{a+1}\right]\frac{V_0 - V}{V} + R T_0 \ln \frac{V}{V_0}$$

$$= F(T, V) - F(T, V_0)\ .$$

Solution 3.9.2

$$2.59 \quad \implies \quad \left(\frac{\partial U}{\partial V}\right)_T = T\left(\frac{\partial p}{\partial T}\right)_V - p\ .$$

That means for the photon gas:

$$\varepsilon(T) = \frac{1}{3}\left(T\frac{d\varepsilon}{dT} - \varepsilon\right) \quad \Longleftrightarrow \quad 4\,\varepsilon(T) = T\frac{d\varepsilon}{dT}$$

$$\implies \quad \varepsilon(T) = \sigma T^4\ .$$

It follows therewith the caloric equation of state:

$$U(T, V) = \sigma V T^4\ .$$

We determine the entropy as follows:

$$T\left(\frac{\partial S}{\partial T}\right)_V = \left(\frac{\partial U}{\partial T}\right)_V = 4\sigma V T^3$$

$$\implies \quad \left(\frac{\partial S}{\partial T}\right)_V = 4\sigma V T^2\ .$$

A first integration yields:

$$S(T, V) = \frac{4}{3}\sigma V T^3 + f(V)\ .$$

We apply now the following Maxwell relation of the free energy:

$$\left(\frac{\partial S}{\partial V}\right)_T = \left(\frac{\partial p}{\partial T}\right)_V = \frac{1}{3}\frac{d}{dT}\left(\sigma\,T^4\right) = \frac{4}{3}\sigma\,T^3$$

$$= \frac{4}{3}\sigma\,T^3 + f'(V)$$

$$\Longrightarrow \quad f(V) = \text{const}.$$

We know therewith the entropy as function of T and V:

$$S(T,V) = \frac{4}{3}\sigma\,T^3\,V + \text{const}.$$

The third law of thermodynamics says that the constant must be zero. We solve for T

$$T = \left(\frac{3}{4\sigma}\right)^{1/3} S^{1/3}\,V^{-1/3}$$

and insert the result into the caloric equation of state:

$$U(S,V) = \left[\frac{3}{4}\left(\frac{3}{4\sigma}\right)^{1/3}\right] V^{-1/3}\,S^{4/3}.$$

That is the internal energy of the photon gas as function of its natural variables S and V.

The free energy is simpler to be calculated:

$$F(T,V) = U(T,V) - T\,S(T,V) = -\frac{1}{3}\sigma\,V\,T^4.$$

For the free enthalpy an especially simple expression comes out:

$$G = F + pV = -\frac{1}{3}\sigma\,V\,T^4 + \frac{1}{3}\sigma\,V\,T^4 = 0.$$

According to the Gibbs-Duhem relation $G = \mu\,N$ we find therewith for the chemical potential of the photon gas:

$$\mu \equiv 0.$$

The enthalpy H is still left:

$$H = U + pV \quad \Longrightarrow \quad H = \frac{4}{3}\sigma\,V\,T^4 = S\,T.$$

It follows with $T = \left(\frac{3p}{\sigma}\right)^{1/4}$:

$$H(S,p) = \left(\frac{3}{\sigma}\right)^{1/4} p^{1/4} S .$$

Solution 3.9.3 Work:

$$\delta W = -F_k \, dx = +k(T) \, x \, dx .$$

Analogy to the gas:

$$\delta W = -p \, dV \implies \quad p \iff -(kx) ,$$
$$V \iff x .$$

For a state change characterized by x and T the free energy F is the proper thermodynamic potential:

$$dF = -S \, dT + k \, x \, dx .$$

For an isothermal elongation of the spring it holds:

$$F(x, T = \text{const}) = \int_{x_0}^{x} kx' \, dx' + F(x_0, T)$$

$$= \frac{1}{2} k \left(x^2 - x_0^2\right) + F(x_0, T) ,$$

$$\Delta F(x, T = \text{const}) = \frac{1}{2} k \left(x^2 - x_0^2\right) .$$

The entropy change results from the Maxwell relation of F:

$$\left(\frac{\partial S}{\partial x}\right)_T = -\left[\frac{\partial}{\partial T}(kx)\right]_x = -x \left(\frac{\partial k}{\partial T}\right)_x$$

$$\implies \quad (\Delta S)_{T=\text{const}} = -\int_{x_0}^{x} x' \frac{dk}{dT} \, dx' = -\frac{1}{2} \frac{dk}{dT} \left(x^2 - x_0^2\right) .$$

We calculate therewith the change of the internal energy:

$$(\Delta U)_T = (\Delta F)_T + T(\Delta S)_T$$

$$= \frac{1}{2} \left(x^2 - x_0^2\right) \left(k - T \frac{dk}{dT}\right) = k \left(x^2 - x_0^2\right) .$$

Solution 3.9.4

1. Work:

$$\delta W = \sigma \, dL \, .$$

When stretching the band, work is executed **on** the system (sign convention!).
Free energy:

$$F(T, L) = -S \, dT + \sigma \, dL \, .$$

Maxwell relation:

$$-\left(\frac{\partial S}{\partial L}\right)_T = \left(\frac{\partial \sigma}{\partial T}\right)_L = \alpha \, .$$

Internal energy:

$$U(T, L) = F(T, L) + T \, S(T, L) \, .$$

We show that U does depend only on T, but not on L:

$$\left(\frac{\partial U}{\partial L}\right)_T = \left(\frac{\partial F}{\partial L}\right)_T + T\left(\frac{\partial S}{\partial L}\right)_T = \sigma - \alpha \, T = 0 \, .$$

2.

$$\left(\frac{\partial S}{\partial L}\right)_T = -\alpha < 0 \, .$$

The entropy of the band decreases when the band is elongated!

3. Required: $(\partial T / \partial L)_S$.
   Chain rule:

$$\left(\frac{\partial T}{\partial L}\right)_S \left(\frac{\partial L}{\partial S}\right)_T \left(\frac{\partial S}{\partial T}\right)_L = -1 \, ,$$

$$C_L = T\left(\frac{\partial S}{\partial T}\right)_L : \quad \text{heat capacity at constant length; always positive}$$

$$\implies \quad \left(\frac{\partial T}{\partial L}\right)_S = \frac{\alpha}{C_L} T > 0 \, .$$

The temperature raises for an adiabatic elongation!

Solution 3.9.5   Entropy:

$$S = S(T\,H) \implies dS = \left(\frac{\partial S}{\partial T}\right)_H dT + \left(\frac{\partial S}{\partial H}\right)_T dH \overset{!}{=} 0$$

$$\implies \left(\frac{\partial T}{\partial H}\right)_S = -\frac{\left(\frac{\partial S}{\partial H}\right)_T}{\left(\frac{\partial S}{\partial T}\right)_H} = -\frac{T}{C_H}\left(\frac{\partial S}{\partial H}\right)_T.$$

Free enthalpy:

$$dG = -S\,dT - m\,dB_0$$

$$B_0 = \mu_0 H\,; \quad m = M\,V\,; \quad V = \text{const}\,, \quad \text{no variable}$$

$$\implies \left(\frac{\partial S}{\partial B_0}\right)_T = \left(\frac{\partial m}{\partial T}\right)_{B_0}$$

$$\implies \left(\frac{\partial S}{\partial H}\right)_T = \mu_0 V \left(\frac{\partial M}{\partial T}\right)_H = -\mu_0 V \frac{C}{T^2} H$$

$$C : \text{ Curie constant}$$

$$\implies \left(\frac{\partial T}{\partial H}\right)_S = \mu_0 V \frac{C\,H}{C_H\,T}.$$

Solution 3.9.6

1. Basic relation:

$$dS = \frac{1}{T}\,dU - \frac{1}{T}Q\,dL = \frac{1}{T}\left(\frac{\partial U}{\partial T}\right)_L dT + \left[\frac{1}{T}\left(\frac{\partial U}{\partial L}\right)_T - \frac{Q}{T}\right] dL\,.$$

Integrability condition for S:

$$\frac{1}{T}\left(\frac{\partial}{\partial L}\left(\frac{\partial U}{\partial T}\right)_L\right)_T \overset{!}{=} -\frac{1}{T^2}\left[\left(\frac{\partial U}{\partial L}\right)_T - Q\right]$$

$$+ \frac{1}{T}\left[\left(\frac{\partial}{\partial T}\left(\frac{\partial U}{\partial L}\right)_T\right)_L - \left(\frac{\partial Q}{\partial T}\right)_L\right].$$

Because of the integrability condition for U the two double derivatives cancel each other and it remains:

$$\left(\frac{\partial U}{\partial L}\right)_T = Q - T\left(\frac{\partial Q}{\partial T}\right)_L = -T^2\left(\frac{\partial}{\partial T}\frac{Q}{T}\right)_L$$

$$\left(\frac{\partial U}{\partial T}\right)_L = C_L(T,L)\,.$$

- **Internal energy**

  We integrate along the path $(T_0, L_0) \to (T, L_0) \to (T, L)$:

  $$U(T, L) = U(T_0, L_0) + \int_{T_0}^{T} dT'\, C_L(T', L_0)$$

  $$- T^2 \left(\frac{\partial}{\partial T} \left(\frac{1}{T} \int_{L_0}^{L} dL'\, Q(T, L') \right) \right)_L .$$

- **Heat capacity**

  $$C_L(T, L) = \left(\frac{\partial U}{\partial T} \right)_L$$

  $$= C_L(T, L_0) - \left(\frac{\partial}{\partial T} T^2 \left(\frac{\partial}{\partial T} \left(\frac{1}{T} \int_{L_0}^{L} dL'\, Q(T, L') \right) \right)_L \right)_L .$$

- **Entropy**

  It follows with the above relations

  $$\left(\frac{\partial S}{\partial T} \right)_L = \frac{1}{T} C_L(T, L)$$

  $$\left(\frac{\partial S}{\partial L} \right)_T = \left[\frac{1}{T} \left(\frac{\partial U}{\partial L} \right)_T - \frac{Q}{T} \right] = - \left(\frac{\partial Q}{\partial T} \right)_L$$

  by integration along the path $(T_0, L_0) \to (T, L_0) \to (T, L)$:

  $$S(T, L) = S(T_0, L_0) + \int_{T_0}^{T} dT' \frac{1}{T'} C_L(T', L_0) - \int_{L_0}^{L} dL' \left(\frac{\partial Q(T, L')}{\partial T} \right)_{L'} .$$

- **Free energy**

  T, L are the 'natural' variables of the free energy:

  $$F = U - TS = F(T, L) .$$

  By combining the results for U and S we get:

  $$F(T, L) = U(T_0, L_0) - TS(T_0, L_0)$$

  $$+ \int_{T_0}^{T} dT' \left(1 - \frac{T}{T'} \right) C_L(T', L_0) + \int_{L_0}^{L} dL'\, Q(T, L') .$$

2.

- **Internal energy**

  With the special ansatz one easily calculates:

  $$\int_{T_0}^{T} dT' \, C_L(T', L_0) = \frac{1}{2}b(T^2 - T_0^2)$$

  $$\int_{L_0}^{L} dL' \, Q(T, L') = \frac{1}{2}aT^2 \, (L - L_0)^2$$

  $$\left(\frac{\partial}{\partial T} \left(\frac{1}{T} \int_{L_0}^{L} dL' \, Q(T, L') \right) \right)_L = \frac{1}{2}a \, (L - L_0)^2 \, .$$

  For the internal energy that leads to:

  $$U(T, L) = U(T_0, L_0) + \frac{1}{2}b(T^2 - T_0^2) - \frac{1}{2}aT^2(L - L_0)^2 \, .$$

- **Heat capacity**

  $$C_L(T, L) = \left(\frac{\partial U}{\partial T} \right)_L = bT - aT(L - L_0)^2 \, .$$

- **Entropy**

  $$\int_{T_0}^{T} dT' \, \frac{1}{T'} C_L(T', L_0) = \int_{T_0}^{T} dT' \, b = b(T - T_0)$$

  $$\left(\frac{\partial}{\partial T} \int_{L_0}^{L} L' \, Q(T, L') \right)_L = 2aT \int_{L_0}^{L} dL' \, (L' - L_0) = aT(L - L_0)^2 \, .$$

  It follows therewith:

  $$S(T, L) = S(T_0, L_0) + b(T - T_0) - aT(L - L_0)^2 \, .$$

- **Free energy**

  $$F = U(T_0, L_0) - TS(T_0, L_0) + \frac{1}{2}aT^2(L - L_0)^2 - \frac{1}{2}b(T - T_0)^2 \, .$$

3. For the thermal expansion coefficient we need:

$$L = \frac{Q}{aT^2} + L_0$$

$$\curvearrowright \left(\frac{\partial L}{\partial T} \right)_Q = -\frac{2Q}{aT^3} \, .$$

Therewith:

$$\alpha = -\frac{2Q}{aLT^3} = -\frac{2}{T}\left(1 - \frac{L_0}{L}\right) .$$

4. Adiabatic-reversible means:

$$S(T_1, L) \overset{!}{=} S(T_2, L_0)$$

and therewith

$$b(T_1 - T_0) - aT_1(L - L_0)^2 = b(T_2 - T_0) .$$

This eventually yields

$$T_2 = T_1\left(1 - \frac{a}{b}(L - L_0)^2\right) .$$

Solution 3.9.7

1.

$$S(T, V) = -\left(\frac{\partial F}{\partial T}\right)_V = N k_B(\alpha + \ln C_0 V) + N k_B \ln C_1 (k_B T)^\alpha .$$

2.

$$p = -\left(\frac{\partial F}{\partial V}\right)_T = \frac{N k_B T}{V} .$$

3.

$$U = F(T, V) + T S(T, V) = N k_B T \alpha .$$

4.

$$C_V = \left(\frac{\partial U}{\partial T}\right)_V = T\left(\frac{\partial S}{\partial T}\right)_V = N k_B \alpha .$$

5.

$$\kappa_T = -\frac{1}{V}\left(\frac{\partial V}{\partial p}\right)_T = -\frac{1}{V}\left[\left(\frac{\partial p}{\partial V}\right)_T\right]^{-1} = \frac{V}{N k_B T} = \frac{1}{p} .$$

Solution 3.9.8   Internal energy **before** the mixing:

$$U = U_1 + U_2 ; \quad U_1 = \frac{3}{2}Nk_B T_1 ; \quad U_2 = \frac{3}{2}Nk_B T_2 .$$

Since no work is needed to be done and no heat transfer takes place, it yields **after** the mixing:

$$U = \frac{3}{2}(2N)k_B T = \frac{3}{2}Nk_B(T_1 + T_2) \implies T = \frac{1}{2}(T_1 + T_2) .$$

By use of the thermal equation of state one gets additionally:

$$U = \frac{3}{2}pV = \frac{3}{2}p_0(V_1 + V_2) = \frac{3}{2}p_0 V \implies p = p_0 .$$

According to (3.43) it holds for the entropy of the ideal gas:

$$S(U, V, N) = Nc + \frac{3}{2}Nk_B \ln \frac{U}{N} + Nk_B \ln \frac{V}{N}$$

$$= Nc + \frac{3}{2}Nk_B \ln \left(\frac{3}{2}k_B T\right) + Nk_B \ln \left(\frac{k_B T}{p}\right) .$$

Entropy of mixing:

$$\Delta S = S(T, p_0, 2N) - S(T_1, p_0, N) - S(T_2, p_0, N)$$

$$= 3Nk_B \ln \left(\frac{3}{2}k_B \frac{1}{2}(T_1 + T_2)\right) + 2Nk_B \ln \left(\frac{k_B \frac{1}{2}(T_1 + T_2)}{p_0}\right)$$

$$-\frac{3}{2}Nk_B \ln \left(\frac{3}{2}k_B T_1\right) - Nk_B \ln \left(\frac{k_B T_1}{p_0}\right)$$

$$-\frac{3}{2}Nk_B \ln \left(\frac{3}{2}k_B T_2\right) - Nk_B \ln \left(\frac{k_B T_2}{p_0}\right)$$

$$= \frac{3}{2}Nk_B \left[\ln \left(\left(\frac{3}{2}k_B\right)^2 \frac{1}{4}(T_1 + T_2)^2\right) - \ln \left(\frac{3}{2}k_B T_1\right) - \ln \left(\frac{3}{2}k_B T_2\right)\right]$$

$$+Nk_B \left[\ln \left(\left(\frac{k_B}{p_0}\right)^2 \frac{1}{4}(T_1 + T_2)^2\right) - \ln \left(\frac{k_B}{p_0}T_1\right) - \ln \left(\frac{k_B}{p_0}T_2\right)\right]$$

$$= \frac{3}{2}Nk_B \ln \left(\frac{1}{4}\frac{(T_1 + T_2)^2}{T_1 \cdot T_2}\right) + Nk_B \ln \left(\frac{1}{4}\frac{(T_1 + T_2)^2}{T_1 \cdot T_2}\right)$$

$$\curvearrowright \Delta S = \frac{5}{2}Nk_B \ln \left(\frac{1}{4}\frac{(T_1 + T_2)^2}{T_1 \cdot T_2}\right) .$$

For the special case $T_1 = T_2$ it obviously results $\Delta S = 0$, i.e., the Gibb's paradox does not appear.

Solution 3.9.9

1.

$$dF = -S\,dT + B_0\,dm = -S\,dT + \mu_0\,V\,H\,dM$$

$$\implies \left(\frac{\partial F}{\partial M}\right)_T = \mu_0\,V\,H\;.$$

Susceptibility:

$$\chi_T = \left(\frac{\partial M}{\partial H}\right)_T = \left[\left(\frac{\partial H}{\partial M}\right)_T\right]^{-1} = \frac{\mu_0\,V}{\left(\frac{\partial^2 F}{\partial M^2}\right)_T}\;.$$

Free energy:

$$\left(\frac{\partial^2 F}{\partial M^2}\right)_T = \frac{\mu_0\,V}{\chi_T}$$

$$\implies \left(\frac{\partial F}{\partial M}\right)_T = \mu_0\,V \int_0^M \chi_T^{-1}(T,M')\,dM' + f(T)$$

$$\implies F(T,M) = F(T,0) + \mu_0\,V \int_0^M dM' \int_0^{M'} dM''\,\chi_T^{-1}(T,M'') + f(T)\,M\;.$$

This is the most general solution!

Special case:

$$\chi_T(T,M) \equiv \chi_T(T) \quad \text{(e.g. Curie law)}$$

$$\implies \chi_T = \frac{M}{H}\;.$$

It follows then:

$$\left(\frac{\partial F}{\partial M}\right)_T = \mu_0\,V\frac{M}{\chi_T} + f(T) \implies f(T) \equiv 0$$

$$\implies F(T,M) = F(T,0) + \mu_0\,V\frac{1}{2}\frac{M^2}{\chi_T}\;.$$

2. Entropy:

$$S \qquad\qquad = -\left(\tfrac{\partial F}{\partial T}\right)_M$$

$$= S(T,0) - M\tfrac{df(T)}{dT} - \mu_0 V \int\limits_0^M dM' \int\limits_0^{M'} dM'' \left(\tfrac{\partial}{\partial T}\chi_T^{-1}(T,M'')\right)_{M''} .$$

The above **special case**:

$$S(T,M) = S(T,0) - \mu_0 V \tfrac{1}{2} M^2 \left(\frac{d}{dT}\chi_T^{-1}(T)\right) .$$

Internal energy:

$$U = F + TS ,$$

$$U(T,M) = U(T,0) + M\left(f(T) - T\frac{df}{dT}\right)$$

$$+\mu_0 V \int\limits_0^M dM' \int\limits_0^{M'} dM'' \left(\chi_T^{-1}(T,M'') - T\frac{\partial}{\partial T}\chi_T^{-1}(T,M'')\right) .$$

Special case:

$$U(T,M) = U(T,0) + \mu_0 V \tfrac{1}{2} M^2 \left(\chi_T^{-1} - T\frac{\partial \chi_T^{-1}}{\partial T}\right) .$$

Thereby:

$$U(T,0) = F(T,0) + TS(T,0) .$$

Solution 3.9.10 We read off from $dF = -S\,dT + B_0\,dm$ the integrability condition:

$$\left(\frac{\partial S}{\partial m}\right)_T = -\left(\frac{\partial B_0}{\partial T}\right)_m = -\mu_0 \left(\frac{\partial H}{\partial T}\right)_m .$$

Curie-Weiss law:

$$M = \frac{C}{T - T_c} H = V m \quad (V = \text{const}) .$$

Heat capacity:

$$\left(\frac{\partial C_m}{\partial m}\right)_T = T\left[\frac{\partial}{\partial m}\left(\frac{\partial S}{\partial T}\right)_m\right]_T = T\left[\frac{\partial}{\partial T}\left(\frac{\partial S}{\partial m}\right)_T\right]_m$$

$$= -\mu_0 T \left(\frac{\partial^2 H}{\partial T^2} \right)_m = 0$$

$$\implies \quad C_m(T, M) \equiv C_m(T) \;.$$

Internal energy:

$$dU = T \, dS + \mu_0 \, V H \, dM$$

$$\implies \quad \left(\frac{\partial U}{\partial T} \right)_M = T \left(\frac{\partial S}{\partial T} \right)_M = C_M = C_m \;,$$

$$\left(\frac{\partial U}{\partial M} \right)_T = T \left(\frac{\partial S}{\partial M} \right)_T + \mu_0 \, V H$$

$$= T V \left(\frac{\partial S}{\partial m} \right)_T + \mu_0 \, V H$$

$$= -\mu_0 \, T V \left(\frac{\partial H}{\partial T} \right)_m + \mu_0 \, V H$$

$$= -\mu_0 \, T V \frac{M}{C} + \mu_0 \, V \frac{M}{C} (T - T_c)$$

$$= -\mu_0 \, V \frac{M}{C} T_c$$

$$\implies \quad U(T, M) = -\mu_0 \, V T_c \frac{M^2}{2C} + g(T) \;.$$

Because of

$$\left(\frac{\partial U}{\partial T} \right)_M = g'(T) = C_m$$

so that finally:

$$U(T, M) = \int_0^T C_m(T') \, dT' - \mu_0 \, V T_c \frac{M^2}{2C} + U_0 \;.$$

Entropy:

$$S(T, M) = \int_0^T \frac{C_m(T')}{T'} \, dT' + f(M) \;,$$

$$\left(\frac{\partial S}{\partial m} \right)_T = \frac{1}{V} \left(\frac{\partial S}{\partial M} \right)_T = -\mu_0 \left(\frac{\partial H}{\partial T} \right)_m = -\frac{\mu_0}{C} M$$

$$\implies \quad f'(M) \overset{!}{=} -\frac{\mu_0 V}{C} M$$

$$\implies \quad S(T,M) = S_0 + \int_0^T \frac{C_m(T')}{T'} \, dT' - \frac{\mu_0 V}{2C} M^2 \ .$$

Free energy:

$$F(T,M) = U(T,M) - T\,S(T,M)$$

$$= F_0 + \int_0^T C_m(T') \left(1 - \frac{T}{T'}\right) dT' + \frac{\mu_0 V}{2C} M^2 \, (T - T_c) \ .$$

Free enthalpy:

$$G = F - m\,B_0 = F - \mu_0 V M H = F - \frac{\mu_0 V}{C} (T - T_c)\, M^2$$

$$= F_0 + \int_0^T C_m(T') \left(1 - \frac{T}{T'}\right) dT' - \frac{\mu_0 V}{2C} M^2 \, (T - T_c)$$

$$\implies \quad G(T, B_0) = F_0 + \int_0^T C_m(T') \left(1 - \frac{T}{T'}\right) dT' - \frac{V C}{2 \mu_0} B_0^2 \frac{1}{T - T_c} \ .$$

Solution 3.9.11

$$H = U + p V = H(S, p) \ ; \quad dH = T\,dS + V\,dp \ ,$$

$$\left(\frac{\partial H}{\partial p}\right)_V = T \left(\frac{\partial S}{\partial p}\right)_V + V \ .$$

Maxwell relation for U:

$$dU = T\,dS - p\,dV \quad \implies \quad \left(\frac{\partial p}{\partial S}\right)_V = - \left(\frac{\partial T}{\partial V}\right)_S$$

$$\implies \quad \left(\frac{\partial H}{\partial p}\right) = V - T \left(\frac{\partial V}{\partial T}\right)_S \ .$$

Solution 3.9.12

1. First law of thermodynamics:

$$dU = \delta Q + \delta W \ .$$

The work is composed of an electrical and a mechanical part:

$$\delta W_e = V E \, dP \quad \text{(the volume } V \text{ is again to be seen as constant,}$$

$$\text{does \textbf{not} belong to the thermodynamic variables),}$$

$$\delta W_m = \tau \, dL \, .$$

Thus we have for reversible state changes:

$$dU = T \, dS + V E \, dP + \tau \, dL \, .$$

The differential of the free enthalpy,

$$G = U - T S - V EP - \tau L \, ,$$

thus reads:

$$dG = -S \, dT - V P \, dE - L \, d\tau \, .$$

It is a total differential from which we get the Maxwell relation

$$V \left(\frac{\partial P}{\partial \tau} \right)_{T,E} = \left(\frac{\partial L}{\partial E} \right)_{T,\tau} \, .$$

2. There are as many thermodynamic potentials as one can create from U by Legendre transformations:

$$U \, ;$$

$$U - T S \, ; \quad U - V P E \, ; \quad U - L \tau$$

$$\Longleftrightarrow \text{transformation in \textbf{one} variable,}$$

$$U - T S - V P E \, ; \quad U - T S - L \tau \, ; \quad U - V P E - L \tau$$

$$\Longleftrightarrow \text{transformation in \textbf{two} variables,}$$

$$U - T S - V P E - L \tau$$

$$\Longleftrightarrow \text{transformation in \textbf{three} variables.}$$

There are therefore altogether **eight** different thermodynamic potentials.
3. Each potential depends on three variables. That leads in each case to three integrability conditions. Altogether that are then twenty four!

Solution 3.9.13

1.

$$U(T,V,N) = F + TS = F - T \left(\frac{\partial F}{\partial T} \right)_{V,N}$$

$$\implies \left(\frac{\partial U}{\partial N} \right)_{T,V} = \left(\frac{\partial F}{\partial N} \right)_{T,V} - T \left[\frac{\partial}{\partial N} \left(\frac{\partial F}{\partial T} \right)_{V,N} \right]_{T,V} .$$

Note that the derivative with respect to the particle number yields the chemical potential μ only then when the differentiated quantity is a thermodynamic potential:

$$\left(\frac{\partial U}{\partial N} \right)_{T,V} \neq \mu ; \quad \text{but:} \quad \left(\frac{\partial U}{\partial N} \right)_{S,V} = \mu(S,V,N) ,$$

$$\left(\frac{\partial F}{\partial N} \right)_{T,V} = \mu(T,V,N) .$$

This means for the above relation:

$$\left(\frac{\partial U}{\partial N} \right)_{T,V} = \mu(T,V,N) - T \left[\frac{\partial}{\partial T} \left(\frac{\partial F}{\partial N} \right)_{T,V} \right]_{V,N}$$

$$\implies \left(\frac{\partial U}{\partial N} \right)_{T,V} - \mu(T,V,N) = -T \left(\frac{\partial \mu}{\partial T} \right)_{V,N} .$$

2.

$$N = N(T,V,\mu)$$

$$\implies dN = \left(\frac{\partial N}{\partial T} \right)_{V,\mu} dT + \left(\frac{\partial N}{\partial V} \right)_{T,\mu} dV + \left(\frac{\partial N}{\partial \mu} \right)_{T,V} d\mu .$$

Take:

$$x = \frac{\mu}{T}$$

$$\implies \left(\frac{\partial N}{\partial T} \right)_{V,x} = \left(\frac{\partial N}{\partial T} \right)_{V,\mu} + 0 + \left(\frac{\partial N}{\partial \mu} \right)_{T,V} \left(\frac{\partial \mu}{\partial T} \right)_{V,x} ,$$

$$\left(\frac{\partial \mu}{\partial T} \right)_{V,x} = \left[\frac{\partial}{\partial T} (Tx) \right]_{V,x} = x = \frac{\mu}{T} .$$

This yields the intermediate result:

$$\left(\frac{\partial N}{\partial T}\right)_{V,x} = \left(\frac{\partial N}{\partial \mu}\right)_{T,V} \left[\left(\frac{\partial N}{\partial T}\right)_{V,\mu} \left(\frac{\partial \mu}{\partial N}\right)_{T,V} + \frac{\mu}{T}\right].$$

Chain rule:

$$\left(\frac{\partial N}{\partial T}\right)_{\mu,V} \left(\frac{\partial T}{\partial \mu}\right)_{N,V} \left(\frac{\partial \mu}{\partial N}\right)_{T,V} = -1$$

$$\implies \left(\frac{\partial N}{\partial T}\right)_{V,x} = \left(\frac{\partial N}{\partial \mu}\right)_{T,V} \left[\frac{\mu}{T} - \left(\frac{\partial \mu}{\partial T}\right)_{N,V}\right]$$

$$\overset{1.}{=} \frac{1}{T}\left(\frac{\partial N}{\partial \mu}\right)_{T,V} \left(\frac{\partial U}{\partial N}\right)_{T,V}$$

3.

$$U = U(T,V,N)$$

$$\implies dU = \left(\frac{\partial U}{\partial T}\right)_{V,N} dT + \left(\frac{\partial U}{\partial V}\right)_{T,N} dV + \left(\frac{\partial U}{\partial N}\right)_{T,V} dN.$$

From that we read off:

$$\left(\frac{\partial U}{\partial T}\right)_{V,x} = \left(\frac{\partial U}{\partial T}\right)_{V,N} + 0 + \left(\frac{\partial U}{\partial N}\right)_{T,V} \left(\frac{\partial N}{\partial T}\right)_{V,x}.$$

After insertion of the result of part 2. the assertion comes out!

Solution 3.9.14

1.

$$dU = T\,dS - p\,dV + B_0\,dm$$

$$dF = -S\,dT - p\,dV + B_0\,dm.$$

2. (a) Integrability relation for F ↷

$$\left(\frac{\partial S}{\partial V}\right)_{T,m} = \left(\frac{\partial p}{\partial T}\right)_{V,m} \longrightarrow \frac{Nk_B}{V}.$$

(b) Integrability condition for F ↷

$$\left(\frac{\partial S}{\partial m}\right)_{T,V} = -\left(\frac{\partial B_0}{\partial T}\right)_{V,m} \longrightarrow -\frac{m}{\alpha V}.$$

(c) With 1.) and 2.a) one finds

$$\left(\frac{\partial U}{\partial V}\right)_{T,m} = T\left(\frac{\partial S}{\partial V}\right)_{T,m} - p = T\left(\frac{\partial p}{\partial T}\right)_{V,m} - p .$$

Because of

$$T\left(\frac{\partial p}{\partial T}\right)_{V,m} = \frac{Nk_B T}{V} = p$$

that means for the ideal paramagnetic gas:

$$\left(\frac{\partial U}{\partial V}\right)_{T,m} = 0 .$$

(d) Wit 1.) and 2.b) it holds now:

$$\left(\frac{\partial U}{\partial m}\right)_{T,V} = T\left(\frac{\partial S}{\partial m}\right)_{T,V} + B_0 = -T\left(\frac{\partial B_0}{\partial T}\right)_{V,m} + B_0 .$$

Because of

$$T\left(\frac{\partial B_0}{\partial T}\right)_{V,m} = T\frac{m}{\alpha V} = B_0$$

it is also here

$$\left(\frac{\partial U}{\partial m}\right)_{T,V} = 0 .$$

3.

- Since the entropy S is a state quantity, the integration path between two points in the state space can be chosen arbitrarily. Here it appears convenient: $(T_0, V_0, m_0) \longrightarrow (T_0, V_0, m) \longrightarrow (T_0, V, m) \longrightarrow (T, V, m)$, where the particle number N is kept constant:

$$S(T, V, m, N) = S(T_0, V_0, m_0, N) + \int_{m_0}^{m} \left(\frac{\partial S}{\partial m'}\right)_{T_0, V_0} dm'$$

$$+ \int_{V_0}^{V} \left(\frac{\partial S}{\partial V'}\right)_{T_0, m} dV' + \int_{T_0}^{T} \left(\frac{\partial S}{\partial T'}\right)_{V,m} dT'$$

$$= -\frac{1}{\alpha V_0} \int_{m_0}^{m} m' \, dm' + Nk_B \int_{V_0}^{V} \frac{dV'}{V'} + C_{V,m} \int_{T_0}^{T} \frac{dT'}{T'} .$$

It thus results for the entropy:

$$S(T,V,m,N) = S(T_0, V_0, m_0, N) - \frac{1}{2\alpha V_0}(m^2 - m_0^2) + Nk_B \ln \frac{V}{V_0} + \frac{3}{2} Nk_B \ln \frac{T}{T_0} \ .$$

- Internal energy:
  Because of

$$\left(\frac{\partial U}{\partial V}\right)_{T,m} = \left(\frac{\partial U}{\partial m}\right)_{T,V} = 0$$

it remains

$$U(T,V,m,N) = \frac{3}{2} Nk_B T + U(T_0, V_0, m_0, N) \ .$$

4. The above result for the entropy does, at first glance, not look as if the homogeneity relation (λ: arbitrary real number)

$$S(T, \lambda V, \lambda m, \lambda N) = \lambda S(T, V, m, N)$$

were fulfilled. But one has to take into consideration that the integration-*constant* $S_0(N) \equiv S(T_0, V_0, m_0, N)$ still depends on the particle number N, which is an extensive variable and has actually been considered during the derivations in part 3.) as constant. The homogeneity is guaranteed if the following requirement can be fulfilled:

$$S_0(\lambda N) - \frac{\lambda^2}{2\alpha V_0} m^2 + \frac{1}{2\alpha V_0} m_0^2 + \lambda Nk_B \ln \frac{\lambda V}{V_0} + \frac{3}{2}\lambda Nk_B \ln \frac{T}{T_0}$$

$$\overset{!}{=} \lambda S_0(N) - \frac{\lambda}{2\alpha V_0} m^2 + \frac{\lambda}{2\alpha V_0} m_0^2 + \lambda Nk_B \ln \frac{V}{V_0} + \frac{3}{2}\lambda Nk_B \ln \frac{T}{T_0} \ .$$

Hence it is to require:

$$\lambda S_0(N) = S_0(\lambda N) - \frac{1}{2\alpha V_0} \lambda(\lambda - 1) m^2 + \frac{1}{2\alpha V_0}(1 - \lambda) m_0^2 + \lambda Nk_B \ln \lambda \ .$$

λ is still free. We choose $\lambda = \frac{N_0}{N}$:

$$S_0(N) = \frac{N}{N_0} S_0(N_0) - \frac{1}{2\alpha V_0}(\frac{N_0}{N} - 1) m^2 + \frac{1}{2\alpha V_0}(\frac{N}{N_0} - 1) m_0^2 + Nk_B \ln \frac{N_0}{N} \ .$$

$S_0(N_0)/N_0 \equiv \gamma$ is a constant. We insert $S_0(N)$ into the expression for the entropy from part 3.):

$$S(T, V, m, N) = N\gamma - \frac{1}{2\alpha V_0}\left(\frac{N_0}{N}-1\right)m^2 + \frac{1}{2\alpha V_0}\left(\frac{N}{N_0}-1\right)m_0^2 + Nk_B \ln\frac{N_0}{N}$$

$$-\frac{1}{2\alpha V_0}\left(m^2 - m_0^2\right) + Nk_B \ln\frac{V}{V_0} + \frac{3}{2}Nk_B \ln\frac{T}{T_0}$$

$$= N\gamma - \frac{1}{2\alpha V_0}\frac{N_0}{N}m^2 + \frac{1}{2\alpha V_0}\frac{N}{N_0}m_0^2$$

$$+ Nk_B \ln\frac{V/N}{V_0/N_0} + \frac{3}{2}Nk_B \ln\frac{T}{T_0} \ .$$

The final result shows that the entropy is indeed extensive:

$$S(T, V, m, N) = N\left(\gamma - \frac{N_0}{2\alpha V_0}\frac{m^2}{N^2} + \frac{1}{2\alpha V_0}\frac{m_0^2}{N_0} + k_B \ln\frac{V/N}{V_0/N_0} + \frac{3}{2}k_B \ln\frac{T}{T_0}\right) \ .$$

Solution 3.9.15

1. **Free energy:**
   According to Exercise 3.9.9 we have:

$$F(T, m) = F(T, 0) + \frac{\mu_0}{2V}\frac{m^2}{\chi_T} \ .$$

This means:

$$F(T, m) = F(T, 0) + \mu_0\frac{T - T_c}{2VC}m^2 \ .$$

Internal energy:
According to Exercise 3.9.9 we have:

$$U(T, m) = U(T, 0) + \frac{\mu_0}{2V}m^2\left(\chi_T^{-1} - T\frac{\partial\chi_T^{-1}}{\partial T}\right) ,$$

$$\chi_T^{-1} - T\frac{\partial\chi_T^{-1}}{\partial T} = \frac{1}{C}(T - T_c) - \frac{T}{C} = -\frac{T_c}{C} \ .$$

This means:

$$U(T, m) = U(T, 0) - \frac{\mu_0 T_c}{2VC}m^2 \ .$$

Entropy:

According to Exercise 3.9.9 we have:

$$S(T, m) = S(T, 0) - \frac{\mu_0}{2V} m^2 \left(\frac{d}{dt} \chi_T^{-1} \right) .$$

This means here:

$$S(T, m) = S(T, 0) - \frac{\mu_0}{2 V C} m^2 .$$

2. **Entropy:**

$$C_m(T, m = 0) = T \left(\frac{\partial S}{\partial T} \right)_{m=0}$$

$$\implies \quad S(T, 0) = \int_0^T \frac{C_m(T', 0)}{T'} \, dT' = \gamma T .$$

With the partial result from 1. we get:

$$S(T, m) = \gamma T - \frac{\mu_0}{2 V C} m^2 \quad \text{(take part 4. into consideration!)}$$

Because of

$$m = \frac{C V}{T - T_c} H$$

it follows immediately:

$$S(T, H) = \gamma T - \frac{1}{2} \mu_0 C V \frac{H^2}{(T - T_c)^2} .$$

Free energy:

$$\left(\frac{\partial F}{\partial T} \right)_m = -S(T, m)$$

$$\implies \quad F(T, 0) = F_0 - \int_0^T \gamma T' \, dT' = F_0 - \frac{1}{2} \gamma T^2 .$$

With the result of part 1. we then get:

$$F(T,m) = F_0 - \frac{1}{2}\gamma T^2 + \mu_0 \frac{T - T_c}{2VC} m^2 \; .$$

Internal energy:

$$C_{m=0} = \left(\frac{\partial U}{\partial T}\right)_{m=0}$$

$$\Longleftrightarrow \quad U(T, m = 0) = \frac{1}{2}\gamma T^2 + U_0 \; .$$

With the result of part 1. we have:

$$U(T,m) = U_0 + \frac{1}{2}\gamma T^2 - \frac{\mu_0 T_c}{2VC} m^2 \; .$$

3. **Heat capacities:**

$$C_m = T\left(\frac{\partial S}{\partial T}\right)_m = \gamma T = C_m(T, m = 0) \; ,$$

$$C_H = T\left(\frac{\partial S}{\partial T}\right)_H = \gamma T + \mu_0 CV \frac{T H^2}{(T - T_c)^3} \; .$$

Because of $T > T_c$ it follows: $C_H \geq C_m$.
    Adiabatic susceptibility:
    According to (2.84) it is:

$$\chi_S = \chi_T \frac{C_m}{C_H} \; .$$

Insertion into the above results leads to:

$$\chi_S(T, H) = \frac{C}{T - T_c} \frac{\gamma T}{\gamma T + \mu_0 CV \frac{T H^2}{(T - T_c)^3}}$$

$$\Longrightarrow \quad \chi_S(T, H) = \frac{C}{T - T_c + \frac{\mu_0 CV}{\gamma} \frac{H^2}{(T - T_c)^2}} \; .$$

4. $T_c = 0 \quad \Longrightarrow \quad$ according to part 2.:

$$S(T, H) = \gamma T - \frac{1}{2}\mu_0 CV \frac{H^2}{T^2} \; .$$

The third law of thermodynamics requires:

$$\lim_{T \to 0} S(T, H) = 0 .$$

Our above result comes for $H \neq 0$ to a contradiction. Thus the Curie law cannot be correct for arbitrarily low temperatures!

We had found in Exercise 3.9.9:

$$S(T, m) = S(T, 0) - \frac{\mu_0}{2V} m^2 \left(\frac{d}{dt} \chi_T^{-1} \right) .$$

If it holds $\chi_T(T, m) \equiv \chi_T(T)$, i.e., $m = V \chi_T H$, then it follows:

$$S(T, m) = S(T, 0) + \frac{\mu_0}{2V} m^2 \frac{1}{\chi_T^2} \frac{d\chi_T}{dT}$$

$$\implies \quad S(T, H) = S(T, 0) + \frac{1}{2} \mu_0 V H^2 \frac{d\chi_T}{dT} .$$

In order to fulfill the third law of thermodynamics we thus have to require

$$\lim_{T \to 0} \frac{d\chi_T}{dT} = 0 \quad \Longleftrightarrow \quad \dot{\chi}_T = \text{const} + \mathcal{O}(T^2) ,$$

i.e., χ_T remains finite for $T \to 0$!

Solution 3.9.16 Because of $T_c = 0$:

$$U(T, m) = U_0 + \frac{1}{2} \gamma T^2 \equiv U(T)$$

(cf. Gay-Lussac experiment for the ideal gas).

1. Isotherm: $0 \to H \quad \implies \quad dU = 0$
   This means:

$$\Delta Q = -\Delta W = -\mu_0 \int_0^H H \, dm ,$$

$$dm = \frac{CV}{T_1} dH$$

$$\implies \quad \Delta Q = -\frac{\mu_0 C V}{2 T_1} H^2 < 0 .$$

heat is dissipated!

2. Adiabatic-reversible $\Longleftrightarrow$ $dS = 0$

$$\Longleftrightarrow \quad S(T_1, H) = S(T_f, 0) \; .$$

We use the result

$$S(T, H) \doteq \gamma \, T - \frac{1}{2} \mu_0 \, C \, V \frac{H^2}{T^2}$$

from part 2. of the preceding exercise.

$$\gamma \, T_1 - \frac{1}{2} \mu_0 \, C \, V \frac{H^2}{T_1^2} \overset{!}{=} \gamma \, T_f \Longleftrightarrow \quad T_f = T_1 - \underbrace{\frac{\mu_0 \, C \, V}{2 \gamma} \frac{H^2}{T_1^2}}_{> 0} < T_1 \; .$$

One therefore achieves a *cooling effect*! Compare this result with part 3. of Exercise 2.9.36!

Solution 3.9.17

1. **Pressure:**

$$p = -\left(\frac{\partial F}{\partial V} \right)_T = -\frac{d \, F_0}{dV} - A \, T \frac{\frac{1}{k_B T} \frac{dE(V)}{dV}}{1 - e^{-\frac{E(V)}{k_B T}}} \; e^{-\frac{E(V)}{k_B T}}$$

$$\Longrightarrow \quad p = -\frac{B}{V_0}(V - V_0) + \frac{A \, E_1}{k_B \, V_0} \, n(T, V) \; .$$

Entropy:

$$S = -\left(\frac{\partial F}{\partial T} \right)_V = -A \, \ln\left(1 - e^{-\frac{E(V)}{k_B T}} \right) + A \, T \frac{\frac{E(V)}{k_B T^2} e^{-\frac{E(V)}{k_B T}}}{1 - e^{-\frac{E(V)}{k_B T}}}$$

$$= -A \, \ln\left(1 - e^{-\frac{E(V)}{k_B T}} \right) + A \, \frac{E(V)}{k_B \, T} \, n(T, V) \; .$$

One easily verifies:

$$e^{\frac{E(V)}{k_B T}} = \frac{n + 1}{n} \; ,$$

$$1 - e^{-\frac{E(V)}{k_B T}} = 1 - \frac{n}{n + 1} = \frac{1}{n + 1} \; ,$$

$$\frac{E(V)}{k_B \, T} = \ln(n + 1) - \ln n \; .$$

The entropy reads therewith:

$$S = A \{(n + 1) \ln(n + 1) - n \ln n\} .$$

Internal energy:

$$U = F + TS = F_0(V) + A T \ln \left(1 - e^{-\frac{E(V)}{k_B T}} \right)$$

$$-A T \ln \left(1 - e^{-\frac{E(V)}{k_B T}} \right) + A \frac{E(V)}{k_B} n(T, V)$$

$$\Longrightarrow \quad U = F_0(V) + A \frac{E(V)}{k_B} n(T, V) .$$

2. From 1. it follows for $p = 0$:

$$\frac{B}{V_0} (V_m - V_0) \overset{!}{=} \frac{A E_1}{k_B V_0} n(T, V_m)$$

$$\Longrightarrow \quad V_m = V(p = 0) = \frac{A E_1}{k_B B} n(T, V_m) + V_0 .$$

This is an implicit conditional equation for V_m, which, for instance, can be iterated:

$$n(T, V_m) = \left\{ \exp \left[\frac{1}{k_B T} \left(E_0 - E_1 \frac{V_m - V_0}{V_0} \right) \right] - 1 \right\}^{-1} .$$

That has to be expanded in powers of E_1 around $E_1 = 0$. Since E_1 appears also as factor, the zeroth order suffices for $n(T, V_m)$:

$$V_m \approx V_0 + \frac{A E_1}{k_B B} n(T, V_0) .$$

Expansion coefficient:

$$\beta = \frac{1}{V} \left(\frac{\partial V}{\partial T} \right)_p .$$

Estimation:

$$\frac{1}{V_m} = \frac{1}{V_0 + \frac{A E_1}{k_B B} n(T, V_m)} \approx \frac{1}{V_0} \left(1 - \frac{A E_1}{k_B B V_0} n(T, V_0) \right) ,$$

$$\frac{\partial V_m}{\partial T} = \frac{A E_1}{k_B B} \frac{\partial n}{\partial T} = \frac{A E_1}{k_B B} \frac{\frac{E(V)}{k_B T^2} e^{\frac{E(V)}{k_B T}}}{\left(e^{\frac{E(V)}{k_B T}} - 1 \right)^2}$$

$$= \frac{A\,E_1}{k_B\,B}\frac{1}{T}n(n+1)\ln\frac{n+1}{n}$$

$$\Longrightarrow\quad \beta \approx \frac{1}{V_0}\frac{A}{B}\frac{E_1}{k_B\,T}n_0\,(n_0+1)\ln\frac{n_0+1}{n_0}\;,$$

$$n_0 = n\,(T,V_0) = \left(e^{\frac{E_0}{k_B\,T}} - 1\right)^{-1}\;.$$

$T \to 0$

$$n_0 \approx e^{-\frac{E_0}{k_B\,T}} \to 0\;,$$

$$n_0 + 1 \to 1\;;\quad \ln\frac{n_0+1}{n_0} \approx -\ln n_0 \approx \frac{E_0}{k_B\,T}$$

$$\Longrightarrow\quad \frac{1}{k_B\,T}n_0\,(n_0+1)\ln\frac{n_0+1}{n_0} \approx \frac{E_0}{(k_B\,T)^2}e^{-\frac{E_0}{k_B\,T}} \xrightarrow[T\to 0]{} 0$$

$$\Longrightarrow\quad \beta(T=0) = 0\;;\quad V_m = V_0\;.$$

$k_B\,T \gg E(V)$

$$n(T,V) \approx \frac{k_B\,T}{E(V)}\;;\quad \frac{\partial n}{\partial T} \approx \frac{k_B}{E(V)} = \frac{k_B}{E_0} + \mathcal{O}(E_1)$$

$$\Longrightarrow\quad \beta \approx \frac{1}{V_0}\frac{A\,E_1}{B\,E_0}\;;\quad V_m \approx V_0 + \frac{A\,E_1}{B\,E_0}T\;.$$

3. Equation (2.65) is a convenient starting point:

$$C_p - C_V = T\left(\frac{\partial p}{\partial T}\right)_V\left(\frac{\partial V}{\partial T}\right)_p\;.$$

It is difficult to keep the pressure constant in our equations:

$$\left(\frac{\partial V}{\partial T}\right)_p = \frac{-1}{\left(\frac{\partial T}{\partial p}\right)_V\left(\frac{\partial p}{\partial V}\right)_T} = -\frac{\left(\frac{\partial p}{\partial T}\right)_V}{\left(\frac{\partial p}{\partial V}\right)_T}\;.$$

This means:

$$C_p - C_V = -T\frac{\left[\left(\frac{\partial p}{\partial T}\right)_V\right]^2}{\left(\frac{\partial p}{\partial V}\right)_T}\;,$$

$$\left(\frac{\partial p}{\partial T}\right)_V = \frac{A E_1}{k_B V_0}\left(\frac{\partial n}{\partial T}\right)_V \approx \frac{A E_1}{k_B V_0}\left(\frac{\partial n_0}{\partial T}\right)_V ,$$

$$\left(\frac{\partial p}{\partial V}\right)_T = -\frac{B}{V_0} + \frac{A E_1}{k_B V_0}\left(\frac{\partial n}{\partial V}\right)_T ,$$

$$\left(\frac{\partial n}{\partial V}\right)_T = \frac{-\frac{1}{k_B T} e^{\frac{E(V)}{k_B T}}}{\left(e^{\frac{E(V)}{k_B T}} - 1\right)^2}\frac{\partial E(V)}{\partial V} ,$$

$$\frac{\partial E(V)}{\partial V} = -\frac{E_1}{V_0}$$

$$\implies \left(\frac{\partial p}{\partial V}\right)_T = -\frac{B}{V_0} + 0(E_1^2)$$

$$\implies C_p - C_V \approx T\frac{A^2 E_1^2}{k_B^2 B V_0}\left(\frac{\partial n_0}{\partial T}\right)_V^2 .$$

Solution 3.9.18

1.

$$U = U(S, V, A) ,$$

$$dU = \delta Q + \delta W ,$$

$$\delta W = \delta W_V + \delta W_A ,$$

$$\delta W_V = -p\, dV ,$$

$$\delta W_A = \sigma\, dA .$$

If the area A is enlarged by dA work is done **on** the system:

$$dU = T\, dS - p\, dV + \sigma\, dA .$$

2. Maxwell relation for dU:

$$\left(\frac{\partial T}{\partial A}\right)_{S,V} = \left(\frac{\partial \sigma}{\partial S}\right)_{V,A} = \left(\frac{\partial \sigma}{\partial T}\right)_{V,A}\left(\frac{\partial T}{\partial S}\right)_{V,A}$$

$$= \frac{\frac{d\sigma}{dT}}{\left(\frac{\partial S}{\partial T}\right)_{V,A}} = \frac{T}{C_{V,A}}\frac{d\sigma}{dT}$$

3. Let us abbreviate

$$\gamma = \frac{\alpha}{T_c\, C_{V,A}} ,$$

then it is to be integrated:

$$\frac{dT}{T} = -\gamma \, dA \quad \Longrightarrow \quad \ln T = -\gamma A + \beta \ .$$

Initial values:

$$\beta = \gamma A_0 + \ln T_0$$

$$\Longrightarrow \quad \ln \frac{T}{T_0} = -\gamma \, (A - A_0) \quad \Longrightarrow \quad T = T_0 \, e^{-\gamma (A - A_0)} \ .$$

The temperature decreases with an adiabatic-isochoric enlargement of the surface!

4.

$$dF = d(U - T \, S) = dU - T \, dS - S \, dT$$

$$\Longrightarrow \quad dF = -S \, dT - p \, dV + \sigma \, dA \ .$$

5. Independent variables: T, V, A:

$$\left(\frac{\partial F}{\partial A} \right)_{T,V} = \sigma(T) \ , \quad \text{independent of } V$$

$$\Longrightarrow \quad F(T, V, A) = \sigma(T) A + F_V(T, V) \ .$$

In addition we have:

$$\frac{\partial}{\partial A} \left(\frac{\partial F}{\partial V} \right)_{T,A} = \frac{\partial}{\partial V} \left(\frac{\partial F}{\partial A} \right)_{T,V} = \frac{\partial}{\partial V} \sigma(T) = 0$$

$$\Longrightarrow \quad \left(\frac{\partial F}{\partial V} \right)_{T,A} = f(T, V) \ , \quad \text{independent of } A$$

$$\Longrightarrow \quad F(T, V, A) = \int^V f(T, V') \, dV' + F_A(T, A) \ .$$

Obviously we can write:

$$F(T, V, A) = F_V(T, V) + F_A(T, A) \ .$$

F_A can be specified explicitly:

$$F_A(T, A) = \sigma(T) A \ .$$

6. Maxwell relation for F:

$$\left(\frac{\partial S}{\partial A}\right)_{T,V} = -\left(\frac{\partial \sigma}{\partial T}\right)_{V,A} = \frac{\alpha}{T_c} > 0 .$$

The entropy S increases when the surface is enlarged!

7. $dU = T\,dS + \sigma\,dA$, if isochoric

$$\implies \left(\frac{\partial U}{\partial A}\right)_{T,V} = T\left(\frac{\partial S}{\partial A}\right)_{T,V} + \sigma \overset{6.}{=} \alpha\frac{T}{T_c} + \alpha\left(1 - \frac{T}{T_c}\right) = \alpha > 0 .$$

8.

$$S = -\left(\frac{\partial F}{\partial T}\right)_{V,A} = S_V(T,V) + S_A(T,A)$$

$$\implies S_A(T,A) = -\left(\frac{\partial F_A}{\partial T}\right)_{V,A} \overset{5.}{=} -A\frac{d\sigma}{dT} = +A\frac{\alpha}{T_c} .$$

$A_1 \to A_2$: isotherm-isochoric $\iff$ $S_V = \text{const}$

$$\Delta Q = T\left(S_A(T,A_2) - S_A(T,A_1)\right) = \alpha\frac{T}{T_c}(A_2 - A_1) ,$$

$\Delta Q > 0$, if $A_2 > A_1$.

9.

$$dG = d(F + p\,V) = -S\,dT + V\,dp + \sigma\,dA .$$

10.

$$\left(\frac{\partial G}{\partial A}\right)_{T,p} = \sigma(T) , \quad \text{independent of } p$$

$$\implies \frac{\partial}{\partial p}\left(\frac{\partial G}{\partial A}\right)_{T,p} = 0 = \frac{\partial}{\partial A}\left(\frac{\partial G}{\partial p}\right)_{T,A}$$

$$\implies \left(\frac{\partial G}{\partial p}\right)_{T,A} = V(T,p) , \quad \text{independent of } A.$$

This means:

$$G(T,p,A) = G_V(T,p) + G_A(T,A) .$$

Surface-part:

$$G_A(T,A) = \sigma(T)A \, ,$$

$$V = \left(\frac{\partial G}{\partial p}\right)_{T,A} = \left(\frac{\partial G_V}{\partial p}\right)_{T,A} \, .$$

Solution 3.9.19

1.

$$\left.\begin{array}{l} G_V^{(1)}(T,p) = M_1\, g_1(T,p) \, , \\ G_A^{(1)}(T,A) = \sigma(T)\, A_1 = \sigma(T)\, 4\pi\, r^2 \, , \end{array}\right\} \quad \text{(drop)}$$

$$G^{(2)}(T,p) = M_2\, g_2(T,p) \qquad \text{(vapor; \textbf{no surface})}$$

$$\implies \quad G(T,p,A) = M_1\, g_1(T,p) + \sigma(T)\, 4\pi\, r^2 + M_2\, g_2(T,p) \, .$$

2. Equilibrium means: $dG = 0$
Since T and p are fixed, only M_1, M_2 and r are variable:

$$M_1 + M_2 = M = \text{const} \quad \implies \quad dM_1 = -dM_2 \, .$$

Part 1. then yields:

$$0 = dG = dM_1\, (g_1 - g_2) + \sigma\, 8\pi\, r\, dr$$

$$\implies \quad g_2 - g_1 = \sigma\, 8\pi\, r\frac{dr}{dM_1} \, .$$

The mass density ρ_1 of the liquid drop,

$$\rho_1 = \frac{M_1}{\frac{4\pi}{3}\, r^3} \, ,$$

is to be considered as constant:

$$\implies \quad M_1 = \rho_1\, \frac{4\pi}{3}\, r^3 \quad \implies \quad \frac{dM_1}{dr} = \rho_1\, 4\pi\, r^2 \, .$$

Therewith the assertion

$$g_2 - g_1 = \frac{2\sigma}{r\,\rho_1}$$

is proven.

3. From the general relation

$$\left(\frac{\partial G}{\partial p}\right)_T = V$$

it follows here:

$$V = V_1 + V_2 = M_1 \left(\frac{\partial g_1}{\partial p}\right)_T + 0 + M_2 \left(\frac{\partial g_2}{\partial p}\right)_T \;.$$

This means obviously:

$$\left(\frac{\partial g_i}{\partial p}\right)_T = \frac{V_i}{M_i} \;;\quad i = 1, 2$$

$$\Longrightarrow \quad \frac{V_2}{M_2} - \frac{V_1}{M_1} = \frac{1}{\rho_2} - \frac{1}{\rho_1} = \left[\frac{\partial}{\partial p}(g_2 - g_1)\right]_T$$

$$= -\frac{2\sigma(T)}{r^2\,\rho_1}\frac{dr}{dp} \;.$$

$\rho_1 \gg \rho_2$:

$$\frac{1}{\rho_2} \approx -\frac{2\,\sigma(T)}{r^2\,\rho_1}\frac{dr}{dp} \;.$$

Vapor = ideal gas:

$$\rho_2 = \frac{M_2}{V_2} = \frac{M_2}{N_2\,k_{\mathrm B}\,T\frac{1}{p}} = \frac{m\,p}{k_{\mathrm B}\,T} \quad (m: \text{ mass of a molecule}) ,$$

$$\Longrightarrow \quad \frac{k_{\mathrm B}\,T}{m\,p} \approx -\frac{2\sigma}{r^2\,\rho_1}\frac{dr}{dp} \quad \Longrightarrow \quad \frac{dp}{p} = \frac{2\,\sigma\,m}{\rho_1\,k_{\mathrm B}\,T}\left(-\frac{dr}{r^2}\right)$$

$$\Longrightarrow \quad \ln p = \frac{2\,\sigma\,m}{\rho_1\,k_{\mathrm B}\,T}\frac{1}{r} + \alpha \;.$$

$p_\infty(T)$: vapor pressure at infinite radius of the drop:

$$\Longrightarrow \quad \alpha = \ln p_\infty$$

$$\Longrightarrow \quad \ln \frac{p}{p_\infty} = \frac{2\,\sigma\,m}{\rho_1\,k_B\,T}\,\frac{1}{r}\,.$$

Vapor pressure of the drop:

$$p(r,T) = p_\infty(T) \exp\left(\frac{2\,m\,\sigma(T)}{\rho_1\,k_B\,T}\,\frac{1}{r}\right)\,.$$

Solution 3.9.20

1. First law of thermodynamics: $dU = \delta Q + \mu_0\,V\,H\,dM$

$$\Longrightarrow \quad C_M = \left(\frac{\partial U}{\partial T}\right)_M ,$$

$$\delta Q = C_M\,dT + \left[\left(\frac{\partial U}{\partial M}\right)_T - \mu_0\,V\,H\right]dM$$

$$\Longrightarrow \quad C_H = \left(\frac{\delta Q}{dT}\right)_H = C_M + \left[\left(\frac{\partial U}{\partial M}\right)_T - \mu_0\,V\,H\right]\left(\frac{\partial M}{\partial T}\right)_H$$

$$\Longrightarrow \quad C_M - C_H = \left[\mu_0\,V\,H - \left(\frac{\partial U}{\partial M}\right)_T\right]\left(\frac{\partial M}{\partial T}\right)_H\,.$$

2.

$$\left(\frac{\partial M}{\partial T}\right)_H = -\frac{C}{T^2}\,H$$

$$\Longrightarrow \quad C_M - C_H = -\frac{\mu_0\,V}{C}\,M^2\,.$$

3a. Maxwell relation of the free energy:

$$dF = -S\,dT + \mu_0\,V\,H\,dM$$

$$\Longrightarrow \quad \left(\frac{\partial S}{\partial M}\right)_T = -\mu_0\,V\left(\frac{\partial H}{\partial T}\right)_M\,.$$

3b. Maxwell relation of the free enthalpy:

$$dG = -S\,dT - \mu_0\,V\,M\,dH$$

$$\Longrightarrow \quad \left(\frac{\partial S}{\partial H}\right)_T = \mu_0\,V\left(\frac{\partial M}{\partial T}\right)_H\,.$$

3c. The assertion follows immediately from 1. for $\delta Q = T \, dS$.

4.

$$C_M - C_H \overset{1.}{=} -T \left(\frac{\partial S}{\partial M} \right)_T \left(\frac{\partial M}{\partial T} \right)_H$$

$$= \mu_0 V T \left(\frac{\partial H}{\partial T} \right)_M \left(\frac{\partial M}{\partial T} \right)_H .$$

5.

$$\left(\frac{\partial H}{\partial T} \right)_M = \frac{M}{C} ,$$

$$dH = \frac{M}{C} dT + \frac{1}{C} (T - T_c) \, dM + 3 b M^2 \, dM$$

$$\implies \left(\frac{\partial M}{\partial T} \right)_H \left[3 b M^2 + \frac{1}{C} (T - T_c) \right] = -\frac{M}{C}$$

$$\implies \left(\frac{\partial M}{\partial T} \right)_H = \frac{-M}{3 b M^2 C + (T - T_c)}$$

$$\implies \dot{C}_M - C_H = \frac{-\mu_0 V T M^2}{3 b M^2 C^2 + C (T - T_c)} .$$

6.

$$\frac{\partial}{\partial M} C_M = \left\{ \frac{\partial}{\partial M} \left[T \left(\frac{\partial S}{\partial T} \right)_M \right] \right\}_T = T \left[\frac{\partial}{\partial T} \left(\frac{\partial S}{\partial M} \right)_T \right]_M$$

$$\overset{(3a.)}{=} T (-\mu_0 V) \left(\frac{\partial^2 H}{\partial T^2} \right)_M = 0 .$$

7.

$$\left(\frac{\partial U}{\partial T} \right)_M = C_M(T) .$$

According to part 3. it is also valid:

$$\left(\frac{\partial U}{\partial M} \right)_T = T \left(\frac{\partial S}{\partial M} \right)_T + \mu_0 V H = -\mu_0 V T \left(\frac{\partial H}{\partial T} \right)_M + \mu_0 V H$$

$$= -\mu_0 V T \frac{M}{C} + \mu_0 V \frac{1}{C} (T - T_c) M + \mu_0 V b M^3$$

$$= \mu_0 V \left(b M^3 - \frac{T_c}{C} M \right) .$$

From that it follows by integration:

$$U(T,M) = \mu_0\, V \left(\frac{1}{4}b\,M^4 - \frac{T_c}{2\,C}M^2\right) + f(T)\,,$$

$$\left(\frac{\partial U}{\partial T}\right)_M = C_M(T) = f'(T)$$

$$\Longrightarrow \quad U(T,M) = \mu_0\, V \left(\frac{1}{4}b\,M^4 - \frac{T_c}{2\,C}M^2\right) + \int_0^T C_M(T')\,dT'\,.$$

Analogously one finds the entropy:

$$\left(\frac{\partial S}{\partial T}\right)_M = \frac{1}{T}C_M(T)\,; \quad \left(\frac{\partial S}{\partial M}\right)_T = -\mu_0\, V \left(\frac{\partial H}{\partial T}\right)_M = -\mu_0\, V \frac{M}{C}$$

$$\Longrightarrow \quad S(T,M) = -\mu_0\, V \frac{M^2}{2\,C} + \int_0^T \frac{C_M(T')}{T'}\,dT' + \underbrace{S(0,0)}_{=0\ (2.82)}\,.$$

That means finally for the free energy:

$$F = U - T\,S = F_0 + \mu_0\, V \frac{1}{2\,C}\,(T - T_c)\,M^2 + \mu_0\, V \frac{1}{4}b\,M^4$$

$$+ \int_0^T C_M(T')\left(1 - \frac{T}{T'}\right)\,dT'\,.$$

8.

$$H = M \left[\frac{1}{C}\,(T - T_c) + b\,M^2\right]\,.$$

$H = 0$ thus possesses the solutions:

a) $M = 0\,,$

b) $M_S = \pm\sqrt{\dfrac{1}{b\,C}(T_c - T)}\,.$

We have for the free energy according to part 7.:

$$F = f(T) + \frac{\mu_0 V}{2C} (T - T_c) M^2 + \frac{1}{4}\mu_0 V b M^4$$

$$\implies \quad F(T, M = 0) = f(T)$$

$$F(T, M = \pm M_S) = f(T) + \frac{\mu_0 V}{2C} (T - T_c) \frac{T_c - T}{bC}$$

$$+ \frac{1}{4}\mu_0 V b \frac{1}{(bC)^2} (T_c - T)^2$$

$$= f(T) - \frac{1}{4} \frac{\mu_0 V}{b C^2} (T_c - T)^2 .$$

So it is:

$$F(T, M = \pm M_S) < F(T, M = 0) .$$

Consequently, the *ferromagnetic* solution $M_S \neq 0$ is the stable one. It exists as **real** solution only for $T \leq T_c$.

9. Magnetic susceptibility:

$$\chi_T = \left(\frac{\partial M}{\partial H}\right)_T = \left[\left(\frac{\partial H}{\partial M}\right)_T\right]^{-1} = \frac{1}{\frac{1}{C}(T - T_c) + 3 b M^2}$$

$$\implies \quad \lim_{H \to 0} \chi_T = \frac{1}{\frac{1}{C}(T - T_c) + 3 b M_S^2} = \frac{C}{2(T_c - T)} .$$

χ_T diverges in the zero-field for $T \to T_c$!

For the difference of the heat capacities we use the result of part 5.:

$$\lim_{H \to 0} (C_M - C_H) = \frac{-\mu_0 V T M_S^2}{3 b M_S^2 C^2 + C(T - T_c)}$$

$$= \frac{-\mu_0 V T \frac{1}{bC}(T_c - T)}{3 b C^2 \frac{1}{bC}(T_c - T) + C(T - T_c)} = -\frac{\mu_0 V}{2 b C^2} T .$$

Solution 3.9.21

1. Maxwell relation for the free enthalpy ($dG = -S\,dT + V\,dp$)

$$\left(\frac{\partial S}{\partial p}\right)_T = -\left(\frac{\partial V}{\partial T}\right)_p .$$

Additionally:

$$\left(\frac{\partial S}{\partial T}\right)_p = \frac{C_p}{T} \implies (dS)_p = \left(C_p \frac{dT}{T}\right)_p.$$

Therewith one calculates:

$$V\beta = \left(\frac{\partial V}{\partial T}\right)_p = -\left(\frac{\partial S}{\partial p}\right)_T$$

$$= -\left(\frac{\partial}{\partial p}\int_0^T (dS)_p\right)_T = -\left(\frac{\partial}{\partial p}\int_0^T \left(C_p \frac{dT'}{T'}\right)_p\right)_T$$

$$= -\int_0^T \left(\frac{\partial C_p}{\partial p}\right)_{T'} \frac{dT'}{T'}$$

$$= -\int_0^T \frac{dT'}{T'}(T')^x \left(a' + b'T + c'T^2 + \dots\right)$$

$$= -\left(\frac{1}{x}T^x a' + \frac{b'}{x+1}T^{x+1} + \frac{c'}{x+2}T^{x+2} + \dots\right)$$

$$= -T^x \left(\frac{a'}{x} + \frac{b'}{x+1}T + \frac{c'}{x+2}T^2 + \dots\right).$$

In the second line we have utilized that, according to the third law, the entropy vanishes at zero temperature. The lower limit of integration therefore does not contribute. The subsequent T-integrations are to be performed along a path with $p = $ const. Furthermore:

$$a' = \frac{d}{dp}a; \quad b' = \frac{d}{dp}b; \quad c' = \frac{d}{dp}c \dots$$

That leads to:

$$\frac{V\beta}{C_p} = -\frac{\frac{a'}{x} + \frac{b'}{x+1}T + \frac{c'}{x+2}T^2 + \dots}{a + bT + cT^2 + \dots}$$

The limiting value

$$\lim_{T\to 0}\frac{V\beta}{C_p} = -\frac{a'}{ax}$$

represents a finite constant.

2. From the $T\,dS$-equation (2.74)

$$T\,dS = C_p\,dT - TV\beta\,dp$$

it follows for an adiabatic process:

$$\left(\frac{dT}{dp}\right)_S = T\frac{V\beta}{C_p}\,.$$

Because of 1.) it is then:

$$\lim_{T\to 0}\left(\frac{dT}{dp}\right)_S = 0\,.$$

Adiabatic relaxation does not lead in the limit $T \to 0$ to a lowering of the temperature, which, in the last analysis, expresses nothing but the unattainability of the absolute zero as a consequence of the third law of thermodynamics.

Section 4.3

Solution 4.3.1

1. Clausius-Clapeyron equation:

$$\frac{dp}{dT} = \frac{Q_M}{T(v_g - v_l)} \approx \frac{Q_M}{Tv_g} \approx \frac{p\cdot Q_M}{RT^2}\,.$$

This means:

$$\frac{dp}{p} = d\ln p \approx \frac{Q_M}{R}\frac{dT}{T^2} \curvearrowright \ln p \approx -\frac{Q_M}{RT} + \text{const.}$$

Thus

$$p(T) \approx \alpha \exp\left(-\frac{Q_M}{RT}\right) \quad (\alpha = \text{const.})$$

2. Thermal expansion coefficient:

$$\beta_{\text{coex}} = \frac{1}{V}\left(\frac{\partial V}{\partial T}\right)_{\text{coex}} \approx \frac{1}{v_g}\left(\frac{\partial v_g}{\partial T}\right)\,.$$

Along the line of coexistence we have

$$v_g = v_g(T, p(T))$$

$$\curvearrowright \left(\frac{\partial v_g}{\partial T}\right)_{\text{coex}} = \left(\frac{\partial v_g}{\partial T}\right)_p + \left(\frac{\partial v_g}{\partial p}\right)_T \left(\frac{\partial p}{\partial T}\right)_{\text{coex}} .$$

The vapor can be seen as ideal gas:

$$\left(\frac{\partial v_g}{\partial T}\right)_p = \frac{R}{p}; \quad \left(\frac{\partial v_g}{\partial p}\right)_T = -\frac{RT}{p^2}; \quad \left(\frac{\partial p}{\partial T}\right)_{\text{coex}} \approx \frac{Q_M}{T v_g} .$$

The last step is justified by the approximate Clausius-Clapeyron equation in part 1). It remains:

$$\beta_{\text{coex}} \approx \frac{1}{v_g} \left(\frac{R}{p} - \frac{RT}{p^2} \cdot \frac{Q_M}{T v_g}\right) = \frac{R}{p v_g} \left(1 - \frac{Q_M}{p v_g}\right) .$$

It follows eventually:

$$\beta_{\text{coex}} \approx \frac{1}{T} \left(1 - \frac{Q_M}{RT}\right) .$$

The first term represents the contribution of the ideal gas. The second term, which arises from the cohesive forces, which come into effect at the phase transition, however, in general dominates. Numeric example:

$$H_2O: \quad Q_M \approx 40\,\text{kJ/mole}$$

$$RT \approx 3\,\text{kJ/mole} \quad \text{at} \quad T = 373\,\text{K} \quad \curvearrowright \quad \beta_{\text{coex}} < 0 .$$

Altogether, a compression takes place with increasing temperature along the curve of coexistence.

Solution 4.3.2

1. Maxwell relation for $G(T, p)$:

$$\left(\frac{\partial S_i}{\partial p}\right)_T = -\left(\frac{\partial V_i}{\partial T}\right)_p = -\frac{\alpha_i}{p} .$$

Integration over the pressure:

$$S_i(T, p) = -\alpha_i \ln \frac{p}{p_0} + f_i(T) .$$

Heat capacities:

$$C_p^{(i)}(T) = T\left(\frac{\partial S_i}{\partial T}\right)_p = T\left(\frac{df_i}{dT}\right)_p \overset{!}{=} C_p(T)$$

$$\curvearrowright f_1'(T) = f_2'(T) \curvearrowright f_1(T) = f_2(T) + \gamma \ .$$

Third law of thermodynamics:

$$f_1(T \to 0) = f_2(T \to 0) = 0 \curvearrowright f_1(T) \equiv f_2(T) \ .$$

This means:

$$S_i(T,p) = -\alpha_i \ln \frac{p}{p_0} + f(T) \ .$$

2. At the line of coexistence we obviously have $S_1 \neq S_2$. Hence, it is about a phase transition of first order. For that the Clausius-Clapeyron equation is valid:

$$\frac{d}{dT} p_{\text{coex}} = \frac{\Delta Q_U}{T(V_1 - V_2)} \ .$$

The 'transformation heat'" ΔQ_U is found by:

$$\Delta Q_U = T\Delta S = -T(\alpha_1 - \alpha_2) \ln \frac{p_{\text{coex}}}{p_0} \ .$$

At the line of coexistence the volume, in addition, exhibits a jump:

$$(V_1 - V_2)_{\text{coex}} = T(\alpha_1 - \alpha_2) \frac{1}{p_{\text{coex}}} \ .$$

This immediately yields the slope of the line of coexistence

$$\frac{d}{dT} p_{\text{coex}} = -\frac{p_{\text{coex}}}{T} \ln \frac{p_{\text{coex}}}{p_0} \ .$$

3. We put $x = T$ and $y = p_{\text{koex}}/p_0$ and have then to solve:

$$\frac{d}{dx} y = -\frac{y}{x} \ln y \ .$$

Rearranged that means:

$$\frac{dy}{y} = d \ln y = -\frac{dx}{x} \ln y \curvearrowright \frac{d \ln y}{\ln y} = -\frac{dx}{x} = -d \ln x \ .$$

That can also be written as follows:

$$d\ln(\ln y) = -d\ln x \quad \curvearrowright \quad d\ln\left(x\cdot\ln y\right) = 0 \quad \curvearrowright \quad x\cdot\ln y = x_0 \ .$$

It is therefore:

$$y = \exp\left(\frac{x_0}{x}\right) \ .$$

If we eventually go back from the substitutions, we recognize that the coexistence-pressure decreases exponentially with increasing temperature:

$$p_{\text{coex}}(T) = p_0 \exp\left(\frac{T_0}{T}\right) \ .$$

The transformation heat is then constant along the line of coexistence:

$$\Delta Q_U = -(\alpha_1 - \alpha_2)\, T_0 \ .$$

Solution 4.3.3

1. The assignment is valid:

$$p \quad \Longleftrightarrow \quad B_0 = \mu_0 H \ ,$$
$$V \quad \Longleftrightarrow \quad -m = -V M \ .$$

Clausius-Clapeyron equation (4.19):

$$\frac{dp}{dT} = \frac{\Delta Q}{T_0 \,\Delta V} \ .$$

This means for the superconductor (Fig. A.16):

$$\Delta Q = T_0 \frac{dB_{0C}}{dT}(-\Delta m) \ ,$$
$$\Delta m = V\,(M_{\text{n}} - M_{\text{s}}) \approx -V M_{\text{s}} = V H_{\text{C}} \ .$$

Fig. A.16

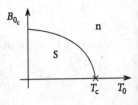

The last step is due to the Meissner-Ochsenfeld effect:

$$\frac{dB_{0C}}{dT} = \mu_0 \frac{dH_C}{dT}$$

$$\implies \quad \Delta Q = -T_0 V \mu_0 \left(H_C \frac{dH_C}{dT} \right)_{T=T_0} .$$

2.

$$G(T, H) = U - TS - \mu_0 V H M ,$$

$$M_n \quad \text{very small} \quad \implies \quad G_n(T, H) \approx G_n(T, 0) ,$$

$$dG = -S \, dT - \mu_0 V M \, dH .$$

Meissner-Ochsenfeld effect:

$$dG_s = -S_s \, dT + \mu_0 V H \, dH .$$

We are interested in the isothermal process:

$$(dG_s)_T = \mu_0 V H \, dH$$

$$\implies \quad G_s(T, H) = G_s(T, 0) + \frac{1}{2} \mu_0 V H^2 .$$

Phase-equilibrium:

$$G_n(T, H_C) \overset{!}{=} G_s(T, H_C) \approx G_n(T, 0) .$$

From that we get the 'stabilization-energy':

$$\Delta G = G_s(T, 0) - G_n(T, 0) \approx G_s(T, 0) - G_s(T, H_C)$$

$$\implies \quad \Delta G = -\frac{1}{2} \mu_0 V H_C^2(T) .$$

3.

$$S_n = -\left(\frac{\partial}{\partial T} G_n(T, H) \right)_H \approx -\left(\frac{\partial}{\partial T} G_n(T, H=0) \right)_{H=0} ,$$

$$S_s = -\left(\frac{\partial}{\partial T} G_s(T, H) \right)_H = -\frac{d}{dT} G_s(T, 0)$$

$$\implies \quad S_n - S_s = -\frac{d}{dT} \Delta G = -\mu_0 V H_C(T) \frac{dH_C(T)}{dT} .$$

This is in accordance with part 1.!
Because of $(dH_C / dT) < 0$:

$$S_n(T) > S_s(T) \ .$$

The superconductor is therefore in a state of higher order. Because of $H_C(T_c) = 0$ it holds at the critical point:

$$S_n(T_c) = S_s(T_c) \ .$$

4. Independently of the values of other parameters it must hold according to the third law of thermodynamics:

$$S_s(T) \xrightarrow[T \to 0]{} 0 \ ; \quad S_n(T) \xrightarrow[T \to 0]{} 0 \ .$$

Since on the other hand it shall be

$$H_C(T) \xrightarrow[T \to 0]{} H_0 \neq 0 \ ,$$

according to part 3. it must be fulfilled

$$\lim_{T \to 0} \frac{dH_C}{dT} = 0 \ ,$$

which is indeed guaranteed by our ansatz for H_C.

5.

$$C_s - C_n = T \left[\frac{\partial}{\partial T} (S_s - S_n) \right]$$

$$= \mu_0 V T \left[\left(\frac{dH_C}{dT} \right)^2 + H_C(T) \frac{d^2 H_C(T)}{dT^2} \right] \ ,$$

$$\frac{dH_C}{dT} = -2 H_0 (1 - \alpha) \frac{T}{T_c^2} - 4 \alpha H_0 \frac{T^3}{T_c^4}$$

$$= -2 H_0 \frac{T}{T_c^2} \left(1 - \alpha + 2 \alpha \frac{T^2}{T_c^2} \right) \ ,$$

$$\left(\frac{dH_C}{dT} \right)^2 = 4 H_0^2 \frac{T^2}{T_c^4} \left(1 - \alpha + 2 \alpha \frac{T^2}{T_c^2} \right)^2 \ ,$$

$$\frac{d^2 H_C}{dT^2} = -2 \frac{H_0}{T_c^2} \left(1 - \alpha + 6 \alpha \frac{T^2}{T_c^2} \right)$$

$$\implies \quad C_s - C_n = \mu_0 \, V \, T \, 2 \frac{H_0^2}{T_c^2} \left[\alpha - 1 + 3 \frac{T^2}{T_c^2} \left(1 - 4\alpha + \alpha^2 \right) \right.$$

$$\left. + 15\,\alpha(1 - \alpha)\frac{T^4}{T_c^4} + 14\,\alpha^2 \frac{T^6}{T_c^6} \right] .$$

The critical point $T = T_c$ is interesting:

$$(C_s - C_n)_{T = T_c} = 4\,\mu_0\,V\frac{H_0^2}{T_c}(1 + \alpha)^2 .$$

6. $\underline{T < T_c}$

$$S_n(T) \neq S_s(T)$$

$$\implies \quad \text{phase transition of first order.}$$

$\underline{T = T_c}$

$$S_n(T_c) = S_s(T_c) ,$$

$$C_n(T_c) \neq C_s(T_c) \quad \text{(finite jump)}$$

$$\implies \quad \text{phase transition of second order.}$$

Solution 4.3.4

$$T = T_c(\varepsilon + 1) .$$

$f(T)$ can be written as follows as function of ε:

$$f(\varepsilon) = a\,T_c(\varepsilon + 1) \ln |T_c\,\varepsilon| + b\,T_c^2(\varepsilon + 1)^2 .$$

The critical exponent is then determined in the following manner:

$$\varphi = \lim_{\varepsilon \to 0} \frac{\ln |f(\varepsilon)|}{\ln |\varepsilon|} = \lim_{\varepsilon \to 0} \frac{\ln |a\,T_c(\varepsilon + 1) \ln |T_c\varepsilon||}{\ln |\varepsilon|}$$

$$= \lim_{\varepsilon \to 0} \frac{\ln |a\,T_c\varepsilon \ln |T_c\,\varepsilon| + a\,T_c \ln |T_c\varepsilon||}{\ln |\varepsilon|} = \lim_{\varepsilon \to 0} \frac{\ln |a\,T_c \ln |T_c\,\varepsilon||}{\ln |\varepsilon|}$$

$$= \lim_{\varepsilon \to 0} \frac{\ln |a\,T_c| + \ln |\ln |T_c\,\varepsilon||}{\ln |\varepsilon|} = \lim_{\varepsilon \to 0} \frac{\ln |\ln T_c + \ln |\varepsilon||}{\ln |\varepsilon|}$$

$$= \lim_{\varepsilon \to 0} \frac{\ln |\ln |\varepsilon||}{\ln |\varepsilon|} = \lim_{\varepsilon \to 0} \frac{\frac{1}{|\ln |\varepsilon||}\frac{1}{|\varepsilon|}}{\frac{1}{|\varepsilon|}} = \lim_{\varepsilon \to 0} \frac{1}{|\ln |\varepsilon||} = 0 .$$

Solution 4.3.5 According to the Ehrenfest-classification, phase transitions of second order are characterized by finite jumps of the second derivatives of the free enthalpy or the free energy:

$$f(\varepsilon) \longrightarrow A_\pm \quad (T \to T_c^{(\pm)}) \;\; ; \quad A_+ \neq A_-$$

$$\Longrightarrow \quad \varphi = \lim_{\varepsilon \to 0} \frac{\ln |f(\varepsilon)|}{\ln |\varepsilon|} = \lim_{\varepsilon \to 0} \frac{\ln |A_\pm|}{\ln |\varepsilon|} = 0 \,.$$

Solution 4.3.6

1.

$$T = T_c(\varepsilon + 1) \quad \Longrightarrow \quad f(\varepsilon) = a\,T_c^5 \,/\, 2(\varepsilon + 1)^{5/2} - b$$

$$\Longrightarrow \quad \varphi = \lim_{\varepsilon \to 0} \frac{\ln |f(\varepsilon)|}{\ln |\varepsilon|} = 0 \,.$$

2.

$$f(\varepsilon) = a\,T_c^2(\varepsilon + 1)^2 + \frac{C}{T_c}\frac{1}{\varepsilon}$$

$$\Longrightarrow \quad \varphi = \lim_{\varepsilon \to 0} \frac{\ln \left|\frac{C}{T_c \varepsilon}\right|}{\ln |\varepsilon|} = -\lim_{\varepsilon \to 0} \frac{\ln |\varepsilon|}{\ln |\varepsilon|} = -1 \,.$$

3.

$$f(\varepsilon) = a\,\sqrt{T_c}\,\sqrt{|\varepsilon|} + d$$

$$\Longrightarrow \quad \varphi = \lim_{\varepsilon \to 0} \frac{\ln |d|}{\ln |\varepsilon|} = 0 \,.$$

Solution 4.3.7 We use (2.82):

$$\chi_T\,(C_H - C_m) = \mu_0\,V\,T\,\beta_H^2; \quad \beta_H = \left(\frac{\partial M}{\partial T}\right)_H$$

$$\Longrightarrow \quad 1 - R = \mu_0\,V\,T\,\beta_H^2\,\chi_T^{-1}\,C_H^{-1} \,.$$

Critical behavior $T \to T_c^{(-)}$:

$$M \sim (-\varepsilon)^\beta \; ; \quad \beta_H^2 \sim (-\varepsilon)^{2\beta - 2} \; ; \quad \chi_T^{-1} \sim (-\varepsilon)^{\gamma'} \; ; \quad C_H^{-1} \sim (-\varepsilon)^{\alpha'}$$

$$\Longrightarrow \quad 1 - R \sim (-\varepsilon)^{2\beta - 2 + \gamma' + \alpha'} \,.$$

From this we read off:

1. $R \neq 1$: The preceding equation is satisfied only if:

$$2\beta - 2 + \gamma' + \alpha' = 0 \quad \Longleftrightarrow \quad \alpha' + 2\beta + \gamma' = 2 .$$

2. $R = 1$: Then the left-hand side of the above equation is zero and can therefore be fulfilled only if

$$2\beta - 2 + \gamma' + \alpha' > 0 \quad \Longleftrightarrow \quad \alpha' + 2\beta + \gamma' > 2 .$$

Solution 4.3.8 A consequence of the scaling hypothesis (4.76) is equation (4.77). In that equation we use

$$\lambda = (\pm\varepsilon)^{-(1/a_\varepsilon)}$$

and obtain with H instead of $B_0 = \mu_0 H$:

$$M(\varepsilon, H) = (\pm\varepsilon)^{(1-a_B)/a_\varepsilon} M\left(\pm 1, (\pm\varepsilon)^{-(a_B/a_\varepsilon)} H\right) .$$

We apply (4.78) and (4.80):

$$\frac{1 - a_B}{a_\varepsilon} = \beta ; \quad \frac{a_B}{a_\varepsilon} = \beta\delta .$$

Therewith it follows immediately the assertion:

$$\frac{M(\varepsilon, H)}{(\pm\varepsilon)^\beta} = M\left(\pm 1, (\pm\varepsilon)^{-\beta\delta} H\right) .$$

One measures the magnetization M for a multitude of external fields H as function of the temperature (or ε). If one then plots

$$\frac{M(\varepsilon, H)}{|\varepsilon|^\beta} \quad \text{versus} \quad \frac{H}{|\varepsilon|^{\beta\delta}} ,$$

then this multitude will reduce to two curves, one for $T < T_c$ and one for $T > T_c$, provided that the scaling hypothesis is valid.

Solution 4.3.9 We apply:

$$(4.78): \quad \beta = \frac{1 - a_B}{a_\varepsilon} ,$$

$$(4.79): \quad \delta = \frac{a_B}{1 - a_B} ,$$

$$(4.81): \quad \gamma = \gamma' = \frac{2a_B - 1}{a_\varepsilon} \,,$$

$$(4.82): \quad \alpha = \alpha' = \frac{2\alpha_\varepsilon - 1}{a_\varepsilon} \,.$$

1. $\gamma(\delta + 1) = (2 - \alpha)(\delta - 1)$
   holds if and only if

$$\frac{2a_B - 1}{a_\varepsilon} \frac{1}{1 - a_B} \overset{!}{=} \frac{1}{a_\varepsilon} \frac{2a_B - 1}{1 - a_B}$$

is fulfilled. That is obviously the case!

2. $\delta = (2 - \alpha + \gamma) / (2 - \alpha - \gamma)$
   is valid if

$$\frac{a_B}{1 - a_B} \overset{!}{=} \frac{2 - \frac{2a_\varepsilon - 1}{a_\varepsilon} + \frac{2a_B - 1}{a_\varepsilon}}{2 - \frac{2a_\varepsilon - 1}{a_\varepsilon} - \frac{2a_B - 1}{a_\varepsilon}}$$

is fulfilled. That is indeed the case:

$$\frac{a_B}{1 - a_B} \overset{!}{=} \frac{2a_\varepsilon - 2a_\varepsilon + 1 + 2a_B - 1}{2a_\varepsilon - 2a_\varepsilon + 1 - 2a_B + 1}$$

$$\Longleftrightarrow \qquad \frac{a_B}{1 - a_B} \overset{!}{=} \frac{2a_B}{2 - 2a_B}$$

Solution 4.3.10

1. We can start with the law of corresponding states (1.19):

$$\left(\pi + \frac{3}{v^2}\right)(3v - 1) = 8t \,,$$

$$p_r = \pi - 1 \,; \quad V_r = v - 1 \,; \quad \varepsilon = t - 1$$

$$\Longrightarrow \quad \left[(1 + p_r) + 3(1 + V_r)^{-2}\right][3(V_r + 1) - 1] = 8(1 + \varepsilon)$$

$$\Longrightarrow \quad \left[4 + 2V_r + V_r^2 + p_r\left(1 + 2V_r + V_r^2\right)\right](3V_r + 2)$$
$$= 8(1 + \varepsilon)\left(1 + 2V_r + V_r^2\right) \,.$$

Arranging a bit the last equation leads to:

$$p_r\left(2 + 7V_r + 8V_r^2 + 3V_r^3\right) = -3V_r^3 + 8\varepsilon\left(1 + 2V_r + V_r^2\right) \,.$$

2. In the critical region all the three quantities p_r, V_r and ε become very small. As a first approximation we can therefore linearize the equation of state from part 1.:

$$p_r \approx 4\varepsilon .$$

In the next step of approximation we insert this again into the equation of state:

$$4\varepsilon \left(2 + 7V_r + 8V_r^2 + 3V_r^3\right) = -3V_r^3 + 8\varepsilon \left(1 + 2V_r + V_r^2\right)$$

$$\implies \quad 0 \approx V_r \left(3V_r^2 + 12\varepsilon + 24V_r\varepsilon + 12\varepsilon V_r^2\right)$$

$$\implies \quad 0 \approx V_r \left(V_r^2 + 8V_r\varepsilon + 4\varepsilon\right) .$$

This equation has the solutions:

$$V_r^{(0)} = 0 ; \quad V_r^{(\pm)} = -4\varepsilon \pm 2\sqrt{-\varepsilon}\sqrt{1 - 4\varepsilon} .$$

$T \overset{>}{\to} T_c \iff \varepsilon \overset{>}{\to} 0 :$
Only $V_r = 0$ can be a solution, because $V_r^{(\pm)}$ are complex.
$T \overset{<}{\to} T_c \iff \varepsilon \overset{<}{\to} 0 :$
We know that the solution $V_r = 0$ is unstable. The reduced volume of the van der Waals gas is therefore:

$$V_r^{(\pm)} = -4\varepsilon \pm 2\sqrt{-\varepsilon}\sqrt{1 - 4\varepsilon} \sim \pm 2\sqrt{-\varepsilon} .$$

3. β determines the behavior of the order parameter (4.52):

$$\frac{\Delta\rho}{2\rho_C} = \frac{1}{2}\frac{\rho^- - \rho^+}{\rho_C} = \frac{V_c}{2}\frac{V^+ - V^-}{V_- V_+}$$

$$= \frac{1}{2}\left(\frac{V_c}{V_-} - \frac{V_c}{V_+}\right) = \frac{1}{2}\left(\frac{1}{V_r^{(-)} + 1} - \frac{1}{V_r^{(+)} + 1}\right)$$

$$\approx \frac{1}{2}\left[1 - V_r^{(-)} - \left(1 - V_r^{(+)}\right)\right] = \frac{1}{2}\left(V_r^{(+)} - V_r^{(-)}\right)$$

$$\implies \quad \frac{\Delta\rho}{2\rho_C} \sim 2\sqrt{-\varepsilon}$$

$$\implies \quad \beta = \frac{1}{2} ; \quad \text{critical amplitude} \quad B = 2 .$$

4. $T = T_c$ means $\varepsilon = 0$. Then the equation of state from part 1. reads:

$$p_r = -3V_r^3 \left(2 + 7V_r + 8V_r^2 + 3V_r^3\right)^{-1} .$$

Expanding for small V_r:

$$p_r = -\frac{3}{2} V_r^3 \left(1 - \frac{7}{2} V_r + \mathcal{O}\left(V_r^2\right) \right) .$$

5. The critical exponent δ is defined in equation (4.57):

$$p_c^{(0)} = n \frac{R T_c}{V_c} = \frac{8}{3} p_c .$$

We have utilized thereby (1.17). Hence it is:

$$\frac{p - p_c}{p_c^{(0)}} = \frac{3}{8} \left(\frac{p}{p_c} - 1 \right) = \frac{3}{8} p_r .$$

Furthermore it holds:

$$\frac{\rho}{\rho_C} - 1 = \frac{V_c}{V} - 1 = \frac{1}{V_r + 1} - 1 = \frac{-V_r'}{V_r + 1}$$
$$= -V_r \left(1 - V_r + 0\left(V_r^2\right) \right) .$$

On the critical isotherm we therefore have, when we exploit part 4. and apply $V_r \to 0$ for $p \to p_c$:

$$\frac{p - p_c}{p_c^{(0)}} \sim \frac{9}{16} \left| \frac{\rho}{\rho_C} - 1 \right|^3 .$$

The comparison with (4.57) yields:

$$\delta = 3 ; \quad D = \frac{9}{16} .$$

6. Compressibility:

$$\kappa_T = -\frac{1}{V} \left(\frac{\partial V}{\partial p} \right)_T = -\frac{1}{V} V_c \left(\frac{\partial V_r}{\partial p} \right)_T ,$$
$$dp_r = d\left(\frac{p}{p_c} - 1 \right) = \frac{1}{p_c} dp ,$$
$$\kappa_T = -\frac{1}{V} \frac{V_c}{p_c} \left(\frac{\partial V_r}{\partial p_r} \right)_T .$$

Normalization factor:

$$\kappa_{T_c}^{(0)} = \frac{1}{p_c^{(0)}} = \frac{V_c}{nRT_c} = \frac{3}{8p_c}.$$

In the last step we utilized again (1.17):

$$\frac{\kappa_T}{\kappa_{T_c}^{(0)}} = -\frac{8}{3}\frac{1}{V_r+1}\left(\frac{\partial V_r}{\partial p_r}\right)_T.$$

According to part 1. we have:

$$\left(\frac{\partial p_r}{\partial V_r}\right)_T = \frac{-9V_r^2 + 16\varepsilon(1+V_r)}{2+7V_r+8V_r^2+3V_r^3}$$

$$-\frac{[-3V_r^3 + 8\varepsilon(1+2V_r+V_r^2)](7+16V_r+9V_r^2)}{(2+7V_r+8V_r^2+3V_r^3)^2}.$$

a) $T \xrightarrow{>} T_c \qquad \rho = \rho_C$, i.e., $V_r = 0$

$$\implies \left(\frac{\partial p_r}{\partial V_r}\right)_{\substack{T \\ V_r=0}} = 8\varepsilon - 14\varepsilon = -6\varepsilon \implies \frac{\kappa_T}{\kappa_{T_c}^{(0)}} = \frac{4}{9}\varepsilon^{-1}.$$

This holds not only for $T \xrightarrow{>} T_c$, but even everywhere along the critical isochore ($V_r = 0$), .

$$\implies \gamma = 1; \quad C = \frac{4}{9}.$$

b) $T \xrightarrow{<} T_c$

In the critical region we have now according to part 2.:

$$V_r^2 \approx -4\varepsilon.$$

This means:

$$\left(\frac{\partial p_r}{\partial V_r}\right)_{\varepsilon \to 0} \approx \frac{1}{2}(36\varepsilon + 16\varepsilon) - \frac{1}{4}56\varepsilon = 12\varepsilon,$$

$$\frac{1}{V_r+1} \xrightarrow{\varepsilon \to 0} 1.$$

It remains:

$$\frac{\kappa_T}{\kappa_{T_c}^{(0)}} \sim -\frac{8}{3}\frac{1}{12\,\varepsilon} = \frac{2}{9}(-\varepsilon)^{-1} \; .$$

so that the comparison with (4.55) yields:

$$\gamma' = 1; \quad C' = \frac{2}{9} = \frac{1}{2}C \; .$$

Solution 4.3.11 Chain rule:

$$\left(\frac{\partial V}{\partial T}\right)_p \left(\frac{\partial T}{\partial p}\right)_V \left(\frac{\partial p}{\partial V}\right)_T = -1$$

$$\Longleftrightarrow \quad (V\beta)\left(\frac{\partial T}{\partial p}\right)_V \left(-\frac{1}{V\kappa_T}\right) = -1$$

$$\Longrightarrow \quad \beta = \kappa_T \left(\frac{\partial p}{\partial T}\right)_V \; .$$

Especially for the van der Waals gas we get:

$$\beta = \kappa_T \left(\frac{nR}{V-nb}\right) \; .$$

The expression in the bracket behaves analytically for $T \to T_c$, so that the critical behavior of β corresponds to that of the compressibility κ_T.

Solution 4.3.12

1. According to (1.28) the equation of state of the Weiss ferromagnet reads:

$$M = M_0 L\left(m\frac{B_0 + \lambda\,\mu_0 M}{k_{\rm B} T}\right) \; ,$$

$$\frac{m\lambda\,\mu_0 M}{k_{\rm B} T} = \frac{M}{M_0}\frac{\frac{N}{V}m^2\lambda\,\mu_0}{k_{\rm B} T} \overset{(1.26)}{=} \widehat{M}\frac{3 k_{\rm B} C\lambda}{k_{\rm B} T} \overset{(1.30)}{=} \widehat{M}\frac{3 T_c}{T} \; .$$

Therewith it follows immediately:

$$\widehat{M} = L\left(b + \frac{3\widehat{M}}{\varepsilon + 1}\right) \; .$$

2. $L(x) = (1/3)x - (1/45)x^3 + 0(x^5)$

$$B_0 = 0 \quad \Longrightarrow \quad b = 0 \,,$$

$$T \leq T_c \quad \Longrightarrow \quad \widehat{M} \text{ very small.}$$

Then:

$$\widehat{M} \approx \frac{\widehat{M}}{\varepsilon + 1} - \frac{3}{5}\frac{\widehat{M}^3}{(\varepsilon + 1)^3}$$

$$\Longrightarrow \quad \frac{\varepsilon}{\varepsilon + 1} \approx -\frac{3}{5}\frac{\widehat{M}^2}{(\varepsilon + 1)^3} \quad \Longrightarrow \quad \widehat{M}^2 \approx -\frac{5}{3}\varepsilon(\varepsilon + 1)^2 \,.$$

Since $(\varepsilon + 1)^2 \to 1$ for $T \to T_c$, we have:

$$\widehat{M} \sim \sqrt{\frac{5}{3}}(-\varepsilon)^{1/2} \,.$$

As for the van der Waals gas we get the critical exponent:

$$\beta = \frac{1}{2} \,.$$

3. Critical isotherm: $T = T_c$; $B_0 \to 0$

$$\Longrightarrow \quad \varepsilon = 0 ; \quad \widehat{M} \quad \text{and} \quad b \quad \text{very small.}$$

This means:

$$\widehat{M} \approx \frac{1}{3}b + \widehat{M} - \frac{1}{45}\left(b + 3\widehat{M}\right)^3$$

$$\Longrightarrow \quad 15b \approx \left(b + 3\widehat{M}\right)^3 \quad \Longleftrightarrow \quad b + 3\widehat{M} \approx (15b)^{1/3}$$

$$\Longrightarrow \quad 3\widehat{M} \approx (15b)^{1/3} - b \approx (15b)^{1/3} \,, \quad \text{da} \quad b \to 0 \,.$$

That yields

$$b \sim \frac{9}{5}\widehat{M}^3$$

and leads to the critical exponent

$$\delta = 3 \,.$$

4.

$$\chi_T = \left(\frac{\partial M}{\partial H}\right)_T = \frac{M_0\,\mu_0\,m}{k_{\rm B}\,T}\left(\frac{\partial \widehat{M}}{\partial b}\right)_{T,b=0} = \frac{3}{\lambda(\varepsilon+1)}\left(\frac{\partial \widehat{M}}{\partial b}\right)_{T,b=0}.$$

In the critical region $\widehat{M}$ is very small:

$$\left.\frac{\partial L}{\partial b}\right|_{b=0} = \left.\frac{\partial x}{\partial b}\left(\frac{1}{3}-\frac{1}{15}x^2\right)\right|_{b=0} + \cdots$$

$$\left.\frac{\partial \widehat{M}}{\partial b}\right|_{b=0} = \left(1 + \frac{3}{\varepsilon+1}\left.\frac{\partial \widehat{M}}{\partial b}\right|_{b=0}\right)\left(\frac{1}{3}-\frac{1}{15}\frac{9\,\widehat{M}^2}{(\varepsilon+1)^2}\right) + \cdots$$

$$\Longrightarrow\quad \left.\frac{\partial \widehat{M}}{\partial b}\right|_{b=0}\cdot\left(1-\frac{1}{\varepsilon+1}+\frac{9}{5}\frac{\widehat{M}^2}{(\varepsilon+1)^3}\right) = \frac{1}{3}\left(1-\frac{9}{5}\frac{\widehat{M}^2}{(\varepsilon+1)^2}\right).$$

$T \to T_{\rm c}$ means $\widehat{M} \to 0$:

$$\left(\frac{\partial \widehat{M}}{\partial b}\right)_{T,b=0} \approx \frac{1}{3}\frac{1}{\frac{\varepsilon}{\varepsilon+1}+\frac{9}{5}\frac{\widehat{M}^2}{(\varepsilon+1)^2}}.$$

a) $T \overset{>}{\to} T_{\rm c}$:

Above $T_{\rm c}$ we have $\widehat{M} \equiv 0$, so that with $(\varepsilon+1) \to 1$ for $T \to T_{\rm c}$ it follows:

$$\left(\frac{\partial \widehat{M}}{\partial b}\right) \sim \frac{1}{3}\varepsilon^{-1}.$$

This means for the susceptibility:

$$\chi_T \sim \frac{1}{\lambda}\varepsilon^{-1} \quad\Longrightarrow\quad \gamma = 1.$$

b) $T \overset{<}{\to} T_{\rm c}$:

According to part 2. we have to now insert $\widehat{M}^2 \sim 5\,/\,3(-\varepsilon)$:

$$\chi_T \sim \frac{1}{2\lambda}(-\varepsilon)^{-1} \quad\Longrightarrow\quad \gamma' = 1.$$

The critical amplitude is equal to that of the van der Waals gas:

$$C' = \frac{1}{2}C.$$

Index